Divide and Prosper

The Heirs of I.G. Farben under Allied Authority 1945–1951

RAYMOND G. STOKES

JEREMY MILLS
PUBLISHING LIMITED

First published by
University of California Press
Berkeley and Los Angeles, California

University of California Press, Ltd.
London, England

This edition published in 2009 by
Hexagon Press
an imprint of
Jeremy Mills Publishing Limited
113 Lidget Street, Lindley, Huddersfield,
West Yorkshire HD3 3JR
www.jeremymillspublishing.co.uk

Copyright © Raymond G. Stokes

ISBN 978-1-906600-15-0

All rights reserved. No part of the contents of this book may be reproduced or transmitted in any form or by any means, electronic or mechanical, including photocopy, recording, or any other information storage or retrieval system, without prior written permission from the publisher.

To my family

CONTENTS

FOREWORD ... vii

NOTES ... xvii

ACKNOWLEDGMENTS ... xxi

INTRODUCTION ... xxiii

PART I. THE BACKGROUND

1. THE RISE AND PRECIPITOUS FALL OF GERMAN *GROSSCHEMIE*, 1860–1945 ... 3

PART II. THE I.G. IN THE INITIAL OCCUPATION PERIOD

2. RADICAL DECONCENTRATION AND ITS RESULTS: THE AMERICAN OCCUPATION AND THE HOECHST GROUP, 1945–1946 ... 37

3. BUSINESS AS USUAL: THE BRITISH AND THE I.G. NIEDERRHEINGRUPPE, 1945–1946 ... 64

4. THE TECHNICAL LIMITS TO EXPLOITATION: THE FRENCH AND THE BADISCHE (BASF), 1945–1948 ... 86

PART III. THE RETURN TO WORLD COMPETITIVENESS: THE I.G. SUCCESSORS, 1947–1951

5. THE CONTINUING CRISIS, JANUARY 1947–JUNE 1948 ... 109

6. CONSOLIDATION OF RECOVERY, MID-1948 TO 1950 ... 136

7. NEW BEGINNINGS AND RESURGENCE,
   1950–1951   164

PART IV. SUMMARY AND CONCLUSIONS

8. FROM COLLAPSE TO COMPETITIVENESS:
   CHANGING FORTUNES AFTER 1945   201

   NOTES   211

   SELECTED BIBLIOGRAPHY   259

   INDEX   287

FOREWORD

2008 marked the twentieth anniversary of the original publication of *Divide and Prosper*, my first book, which was a heavily revised version of my PhD dissertation.[1] It tells the story of the forced break-up in the aftermath of World War II of one of the largest firms in the world, the German I.G. Farbenindustrie AG, and the (re)founding of its successor companies, BASF, Bayer, and Hoechst, each of which has been a dominant force in the post-war world chemical industry. Despite the fact that the book has been out of print for more than half of the period between original publication and this new edition, it remains a heavily cited source in scholarship on the Allied occupation of Germany after World War II and the beginnings of West Germany's "economic miracle". Given this, and given that the first edition appears to be available for purchase from internet booksellers only at exorbitant prices, it is a pleasure to see it back in print with Jeremy Mills Publishing, and I am grateful to Jeremy Mills for providing me this opportunity, and also to Peter Morris of the Science Museum in London, a friend and colleague for a long time, who drew Jeremy Mills's attention to the possibility of reprinting *Divide and Prosper*.

What I would like to do in this brief foreword to the new edition is to discuss the context within which I began researching and writing the book; to examine how it has been received and/or extended; and to suggest some related areas of research which still require scholarly investigation.

THE CONTEXT OF RESEARCHING AND WRITING

There were basically four reasons for me choosing to write about the post-1945 break-up of the giant German chemical firm, I.G.

Farbenindustrie AG – one personal, one pragmatic, one opportunistic, and one both conceptually challenging and with considerable scholarly interest.

My personal interest in choosing to begin research in the early 1980s on the subject of the break-up of the I.G. was stimulated by two books that I had read in the late 1970s. The first, the novel *Gravity's Rainbow* by Thomas Pynchon, featured I.G. Farben as a key and shadowy character, but also depicted Allied occupied Germany as a "zone" in which traditional power relationships had ceased to exist, while new ones had yet to be forged.[2] I was fascinated by this idea, which, however, also seemed to be challenged by the other book, *The Crime and Punishment of I.G. Farben* by Joseph Borkin.[3] Borkin was a lawyer who had served in the U.S. Justice Department under Thurman Arnold in the antitrust division in the 1930s, investigating, among other things, the activities of I.G. Farben in the United States. He was also involved in wartime agitation targeting I.G. Farben as a disruptive force, not only economically, but also politically.[4] Echoing at times the conclusions of Josiah DuBois,[5] the prosecutor of I.G. Farben executives at one of the American-led trials of alleged war criminals in Nuremberg after the war,[6] Borkin essentially argued that the firm and its managers had committed crimes during, but avoided punishment after World War II mainly owing to the *persistence* of traditional power relationships (including those between companies in Germany and Allied countries which had pre-dated the war) during and beyond the Allied occupation. The occupation, he suggested, thus only *appeared* to provide an opportunity to challenge those traditional relationships. Each of the major successors of I.G. Farben was, by the time he wrote his book, extremely successful as well, and each was much larger than the original firm had ever been.

Investigating this puzzle of conflicting interpretations – which essentially involved the quintessential historical problem of the relationship between continuity and change – was something I decided to do. Borkin's book, after all, was more a lawyer's brief than a work of scholarship, and most of it was concerned with the I.G. in the Nazi period. Pynchon's book was a work of creative reinterpretation, not bound by the rules of historical scholarship. In retrospect, I was wise to avoid dealing with the I.G. in the earlier period since, after starting my research in 1983, I soon became aware of recent and extensive work on the firm in the Nazi period by Peter Morris and Peter Hayes.[7] In contrast, the only scholarly treatment of the subject of its break-up after the war was a fairly brief article by Hans-Dieter Kreikamp which was based primarily on German sources.[8]

But I also had a strong pragmatic reason for choosing to study the I.G.'s fate in the Allied occupation period: limited German language skills, at least at that time. I had studied French intensively as an undergraduate, but had only started to learn German as a graduate student. Although my German improved substantially as a result of a Goethe Institute course in Mannheim in 1981, it still required polishing. It seemed a wise idea to choose a topic which allowed me to use the languages in which I was research-ready, English and French, while working at getting my German in order, which I was able to do during a two-year period at the Institute for Contemporary History in Mainz between 1983 and 1985.

There was also an element of opportunism in the choice of western Germany during the Allied occupation period. Under the "thirty-year rule", to which all four of the countries involved – West Germany, the United States, Great Britain, and France – subscribed, official documents relating to the period of military occupation, i.e. 1945-1949, were in the process of becoming available for scholarly research in the late 1970s, and there were vast quantities of them. It was an opportunity to get in on the ground floor with archival material previously unexamined by historians.

My final reason for choosing this subject was perhaps the most convincing in intellectual terms. It so happened that each of the major successors to I.G. Farben, i.e. Hoechst, Bayer, and BASF, was located in a different western zone of occupation, i.e. the American, the British, and the French respectively. It was, I thought, a rare opportunity for investigation of the effects of differing policy regimes on relatively similar companies of similar size, all of which had recently been part of a single company. As is often the case in historical research, things turned out to be more complicated than that: all of the companies involved, after all, had had a long "pre-I.G." existence; they were not as similar to one another as it had appeared to me in the early days; and there was a dynamic in the power relationships among the western occupying powers that led to policy regimes converging over time. But, especially for the early occupation period, the original conceptualisation worked quite well.

The other aspect of the research project which was of scholarly and conceptual interest was the opportunity to use government documents from all four countries involved, and to combine official documents with material from corporate archives, allowing investigation and portrayal of multiple points of perspective about the same issues. Again, the real world differed somewhat from the ideal one I had constructed early on in the project, in particular in relation to company archives: BASF at that time refused me entry, as did Hoechst. Both have long since radically revised their policy in regard to access for scholars, but even at the time of writing *Divide and Prosper* they did provide me with some documents, and Hoechst in particular had published facsimile editions of key documents on the American occupation.[9] There were also materials written by or about the companies in the French and American occupation records as well as in other corporate archives. Bayer had by far the best materials on the period, which were made available to me virtually without restriction, and the British case was thus the one where the triangular "company-occupier government-occupied government" relationship could be investigated most closely.

Following two years of research in Germany, France, Britain, and the United States, I began writing intensively during 1985/6, defending the PhD dissertation successfully in spring of 1986. I was in an unusually fortunate position in three respects. First, I had a fellowship for the remainder of that calendar year (the fellowship ran until December, but only under the condition that I did not submit the final version of the dissertation beforehand), and I used this time to revise the version I had defended fundamentally (there would be

Foreword xi

another major revision later). Second, and partly because of the first point, I had a book contract from the University of California Press before formal submission of the dissertation or award of the doctorate. Finally, again related to the first two, the dissertation was cited (although not by title) in a major work on business in post-war West Germany published even before the dissertation was submitted.[10]

The time afforded by the need to delay submission of the final dissertation enabled me to work towards developing what is I think an unusual approach to a subject of this sort. I have already mentioned the synergies developed through looking at both official governmental and corporate archival and printed published materials, so in a way I am not sure if what I am describing here is a cause or an effect. But what attracted me to the subject, what I sought out in the source base, and what I wrote about in the end is not really business history, nor economic history, nor political history, nor history of technology *per se*, but rather an amalgam of all of these approaches to the historical profession. This approach – for want of a better phrase, a political economic one – might be criticised by anyone from any one of these areas of the profession. But the risk of not satisfying the "purist" historian in any of the specialist areas is, I hope, offset by the complementary evidential and explanatory power of the overlay of approaches. On the original dust jacket Jeffrey Sturchio agreed in a sense, commenting that this is "an important book at the intersection of German history, business history, and the history of technology", while Peter Hayes commented that "the author deftly demonstrates the mutually illuminating skills of a political and an economic historian".

RECEPTION AND EXTENSION

The citation of the dissertation rather than of the book by Berghahn in 1986 (that I just mentioned) is a rare one since the book itself appeared just over a year after the formal submission of the dissertation in late autumn 1986 for graduation in December of that year. Following its official appearance in late 1988 (the book was actually a physical fact in summer 1988), *Divide and Prosper* itself has become the most frequently cited and well known of my publications. The reviews of it were generally positive. Indeed, aside from some reviewers asking for more or less detail on particular issues, most agreed in spirit at least with Peter Becker's comment in the *American Historical Review* that it makes "a valuable contribution to economic history that no one interested in German history ought to miss".[11] Mark Roseman echoed this sentiment in *German History*, saying that it "is a book which deserves a wide readership and which will hold all the readers it gets".[12]

It has been particularly gratifying that this more general readership has indeed been achieved, as indicated by fairly brisk sales of the first edition that clearly went far beyond research libraries, and as evidenced by citations in scholarly literature. Not only historians of German business, technology, and the economy, but also more general historians have made use of the research findings of *Divide and Prosper*. It also figures in even more general diplomatic/international historical accounts, for instance in Carolyn Woods Eisenberg's prizewinning book, *Drawing the Line*.[13] I hope that this new edition will broaden the readership still further.

While the findings of *Divide and Prosper* have been largely accepted, there were of course many questions remaining to be examined in its wake, and some of them have been taken up by other historians (as well as myself in further research). One strand of inquiry involves other firms and industries and their experience in the occupation. The uptake here has been somewhat disappointing, although there are some treatments of the occupation period in the context of longer-term examinations of industrial and company history.[14] Another strand of inquiry has involved the chemical industry in the zone of occupation "missing" from *Divide and Prosper*, i.e. the Soviet zone. Research on this area was hindered prior to 1989 by East German policymakers, but it has flourished since then.[15] Another related area is that of technology transfer and innovation during and after the occupation. Again, significant strides have been made here in the aftermath of the publication of *Divide and Prosper*.[16] One additional

Foreword   *xiii*

area has involved exploring the longer-term development of the companies dealt with in the book.[17]

Two of the most intriguing areas are ones in which I had already developed an interest by the time *Divide and Prosper* came out, but which, owing to other research projects, I have only been able to hint at in my own work.[18] For one thing, this book, along with much of my subsequent writing, has focused on the producer side of the chemical (and some other) industries. But one of the most important developments in Germany and elsewhere has involved the consumer side of the chemical industry, in particular the explosion in the consumption of thermoplastics. Jeffrey Meikle's consideration of *American Plastic* was a model in this regard, considering the linkage between production, technology, design, consumption, and, to a lesser degree, disposal.[19] It is gratifying to see that, recently, books building in part on my work in *Divide and Prosper*, and other publications such as Meikle's, have appeared, one relating to West Germany and one to East Germany. Andrea Westermann looks at the West German history of polyvinyl chloride (PVC), one of the first and most important plastics in the post-war period, from a number of different angles, including the role of research and development in the production process, the links between the raw plastics producers and the plastics processors and designers, and cultures of consumption of plastics as well as environmental issues, about which more shortly.[20] Eli Rubin has done something very similar for the "other" Germany in his book on plastics in East Germany, although his focus is more on the cultural and political meanings of plastics and their deployment in industry and society and less on environmental issues.[21]

The second of these intriguing areas is even less well developed, i.e. the environmental history of the chemical industry in Germany, although there is one sustained historical treatment of it so far, that by Westermann. She addresses this in a number of dimensions, looking at PVC as a carcinogen in the workplace, at issues raised by PVC in relation to the democratic oversight of risks from science and technology, and at PVC as a contributor to growing problems in waste disposal. The links between the environmental history of the chemical industry, the history of public health and that of waste disposal and management, the history of occupational health and safety, and the history of regulation are all touched upon here, although clearly much remains to be done.

A RESEARCH AGENDA

Divide and Prosper was, I believe still, a pioneering work in many respects. It was, to my knowledge, the first scholarly work to examine events in all three western zones of occupation in post-World War II Germany based on a wide range of newly released official documents and newly accessible corporate archives. It provided a partial explanation, at least, for the beginnings of the West German "economic miracle" by indicating the process by which a key industrial sector restructured for competitiveness in the 1950s and beyond. And, as I have already indicated, it also featured a novel approach, integrating economic and business history and, to a lesser degree, history of technology on the one hand with political history on the other, something rare in the historiography of post-war occupied Germany, at least when it was first published.[22]

As noted, a number of aspects suggested (or sometimes simply overlooked) in *Divide and Prosper* have been explored by a range of scholars in the two decades since its initial publication. Nevertheless, there is still a considerable amount of scholarly work to be done to explore some of the questions raised in my book, and, although some of it is underway already, much remains to be undertaken. I will restrict my suggestions here to three areas.

One set of issues concerns the relationship between occupation policy, long-term structural changes in the German economy, and economic and industrial development in relation to firms in a range of industries. To date, large-scale industry, for the most part restricted to heavy industry and chemicals, has been virtually the sole focus of scholarship. Other sectors, and small and medium sized enterprise, should also be examined. Here I note that there is some suggestion of a movement in this direction by scholars, in particular in the consideration of industrial and company planning and practice during the National Socialist period and its longer-term implications.[23]

My second and third suggestions involve extending still further the intriguing developments I noted above. Plastics were and remain part of the political culture of major industrialised countries in the period since World War II, as Meikle and Westermann remind us, but they are also central to material culture and the ways in which it has changed through time. Thus, further exploration of the links between producer, designer, processor, and consumer over time is needed, and not just in relation to one plastic, PVC, but with regard to others as well. What influence did varying materials have on design possibilities, not just in relation to plastic artefacts, but also to the design of other materials and products? What impacts did these developments in turn have on packaging and logistics, and eventually on the consumer? How have all of these developments affected, and been affected by, political, social, and economic trends?

Finally, I would suggest that issues involving the connection between the chemical industry and the environment have only just begun to be explored, but that they are extremely important. How aware was the industry of what the implications of its production processes and the waste generated by them were? When did perceptions change within industry? A closer examination of the development of political backlash against the industry is also worth undertaking, not least since this was a major factor in spurring it (in Germany especially) towards "high chem" production[24] and towards repositioning itself, through a variety of new products and processes, as part of the solution rather than one of the main sources of the problem in regard to the environment.[25] The connection between packaging, much of which was generated by the chemical industry, and waste disposal, which became an enormous problem in part because the packaging in particular generated massive growth in volume of waste, would be particularly useful to explore in terms of this last point. What role have the industry and its companies played in spearheading efforts to solve the challenges of reclaiming discarded plastics through recycling, thus transforming "waste" into something useful? To what extent has this – i.e. the development of plastics recycling technologies which might be exported to other countries – formed part of (West) German industrial policy?[26]

Much remains to be done, but what follows below was a start along this road. I am very pleased that it is at last available again in print, and hope that all readers, whether new or returning to it, will find it of use and interest.

Raymond G. Stokes
Centre for Business History
University of Glasgow
August 2008

NOTES

FOREWORD

1. Raymond G. Stokes, "Recovery and resurgence in the West German chemical industry, 1945-1951" (PhD dissertation, The Ohio State University, 1986); *Divide and Prosper: The Heirs of I.G. Farben Under Allied Authority, 1945-1951* (Berkeley: University of California Press, 1988).

2. Thomas Pynchon, *Gravity's Rainbow* (New York: Viking, 1973).

3. Joseph Borkin, *The Crime and Punishment of I.G. Farben* (New York: The Free Press, 1978).

4. Joseph Borkin and Charles Welsh, *Germany's Master Plan: The Story of Industrial Offensive* (New York: Duell, Sloan, and Pearce, 1943).

5. Cf. Josiah DuBois, *The Devil's Chemists: 24 Conspirators of the International Farben Cartel Who Manufacture Wars* (Boston: Beacon Press, 1952).

6. The most famous trial at Nuremberg, involving primarily political and military officials, was an Allied four-power trial. Subsequently, the Americans carried out a further twelve sets of trials focusing on various other groups (e.g. lawyers, doctors, industrialists). The I.G. Farben trial, trial VI, took place in 1947 and 1948. See *Trials of War Criminals Before the Nuernberg Military Tribunals*, vol. VII (Washington: U.S. Government Printing Office, 1953) (available at http://www.mazal.org/archive/nmt/07/NMT07-C001.htm [viewed 18 August 2008]).

7. Peter Morris, "The Development of acetylene chemistry and synthetic rubber by I.G. Farbenindustrie Aktiengesellschaft, 1926-1945" (PhD dissertation, University of Oxford, 1982); Peter Hayes, "The *Gleichschaltung* of I.G. Farben" (PhD dissertation, Yale University, 1982). Hayes subsequently heavily revised and rewrote his dissertation, which appeared as *Industry and Ideology: I.G. Farben in the Nazi Era* (Cambridge: Cambridge University Press, 1987; 2nd edition, 2001).

8. Hans-Dieter Kreikamp, "Die Entflechtung der I.G. Farbenindustrie AG und die Gründung der Nachfolgegesellschaften," *Vierteljahreshefte für Zeitgeschichte* 25 (1977): 220-251.

9. Volumes 48-50 of the *Dokumente aus Hoechst-Archiven: Beiträge zur Geschichte der chemischen Industrie* put out by the company in 1976, 1978 and 1978 respectively, deal with the American occupation and the negotiations to break up I.G. Farben and to create Hoechst as one of its successors.

10. Volker Berghahn, *The Americanisation of West German Industry, 1945-1973* (Cambridge/New York: Cambridge University Press, 1986), p. 93, n. 76.

11. Peter W. Becker, Review in *The American Historical Review*, Vol. 95, No. 5. (December 1990), p. 1570.

12. Mark Roseman, Review in *German History*, Vol. 8, No. 2 (June 1990), p. 250.

13. Carolyn Eisenberg, *Drawing the Line: The American Decision to Divide Germany, 1944-1949* (Cambridge: Cambridge University Press, 1996).

14. See, for instance, Paul Erker, *Competition and Growth: A Contemporary History of the Continental AG* (Dusseldorf: Econ, 1996); Rainer Karlsch and Raymond G. Stokes, *Faktor Öl: Die Mineralölwirtschaft in Deutschland 1859 bis 1974* (Munich: Beck, 2003); Paul Erker and Bernhard Lorentz, *Chemie und Politik: Die Geschichte der Chemischen Werke Hüls 1938-1980* (Munich: Beck, 2003); Isabel Warner, *Steel and Sovereignty: The Deconcentration of the West German Steel Industry, 1949-54* (Mainz: Philipp von Zabern, 1996).

15. See, for example, Hermann-J. Rupieper, Friederike Sattler, Georg Wagner-Kyora, eds., *Die mitteldeutsche Chemieindustrie und ihre Arbeiter im 20. Jahrhundert* (Halle: mdv, 2005); Rainer Karlsch, "Capacity losses, reconstruction and unfinished modernization: The chemical industry in the Soviet Zone of Occupation (SBZ)/GDR, 1945-1965," pp. 367-405 in John E. Lesch, ed., *The German Chemical Industry in the Twentieth Century* (Amsterdam: Kluwer Academic Publishers, 2000); and Karlsch's overview of the history of the Buna-Werke, pp. 11-43 of Karlsch and Raymond G. Stokes, *The Chemistry Must Be Right: The Privatization of Buna Sow Leuna Olefinverbund GmbH, 1990-2000* (Leipzig: Edition Leipzig, 2001).

16. See, for instance, Burghard Ciesla and Matthias Judt, eds., *Technology Transfer Out of Germany After 1945* (Amsterdam: Harwood Academic Publishers, 1996); Rolf Petri, ed., *Technologietransfer aus der deutschen Chemieindustrie (1925-1960)* (Berlin: Duncker & Humblot, 2004).

17. See, for instance, Raymond G. Stokes, *Opting For Oil* (Cambridge: Cambridge University Press, 1994); Werner Abelshauser and others, *German Industry and Global Enterprise: BASF: The History of a Company* (Cambridge: Cambridge University Press, 2004); Patrick Kleedehn,

Die Rückkehr auf den Weltmarkt: Die Internationalisierung der Bayer AG Leverkusen nach dem Zweiten Weltkrieg bis zum Jahre 1961 (Stuttgart: Franz Steiner, 2007).

18. Raymond G. Stokes, "Gravity and the rainbow makers: Some thoughts on the trajectory of the German chemical industry in the twentieth century," pp. 441-449 in Lesch, ed., *The German Chemical Industry in the Twentieth Century*; Raymond G. Stokes, "Plastics and the new society: The German Democratic Republic in the 1950s and 1960s," pp. 65-80 in Susan E. Reid and David Crowley, eds., *Style and Socialism: Modernity and Material Culture in Post-War Eastern Europe* (Oxford: Berg, 2000).

19. Jeffrey L. Meikle, *American Plastic: A Cultural History* (New Brunswick, NJ: Rutgers University Press, 1995).

20. Andrea Westermann, *Plastik und politische Kultur in Westdeutschland* (Zurich: Chronos, 2007).

21. Eli Rubin, *Synthetic Socialism: Plastics and Dictatorship in the German Democratic Republic* (Chapel Hill, NC: University of North Carolina Press, 2009).

22. The one major exception, of course, was Berghahn's *Americanisation of West German Industry*, but the political-economic approach (still rarely involving technology or the internal workings of the firm) to the history of the occupation period and early Federal Republic of Germany has become less rare more recently. See, for instance, James C. Van Hook, *Rebuilding Germany: The Creation of the Social Market Economy, 1945-1957* (Cambridge: Cambridge University Press, 2004); Ralf Ptak, *Vom Ordoliberalismus zur Sozialen Marktwirtscahft. Stationen des Neoliberalismus in Deutschland* (Opladen: Leske + Budrich, 2004); Reinhard Neebe, *Weichenstellung für die Globalisierung* (Cologne: Böhlau, 2004).

23. On planning and practice in the National Socialist period, see, for instance, Christoph Buchheim and Jonas Scherner, "The role of private property in Nazi Germany: The case of industry," *Journal of Economic History* 66 (2006): 390-416; on the importance of business rather than political issues in business decision-making during the Nazi period, especially in small and medium-sized enterprise, Michael C. Schneider, *Unternehmensstrategien zwischen Weltwirtschaftskrise und Kriegwirtschaft. Chemnitzer Maschinenbauindustrie in der NS-Zeit 1933-1945* (Essen: Klartext, 2005).

24. See Christopher S. Allen, "Political consequences of change: The chemical industry," pp. 157-184 in Peter Katzenstein, ed., *Industry and Politics in West Germany: Toward the Third Republic* (Ithaca, NY: Cornell University Press, 1989). See especially p. 167.

25. See, for instance, various essays in Monica J. Casper, ed., *Synthetic Planet* (London: Routledge, 2003).

26. Answering some of these questions is one aspect of a project I am directing through 2010 on "Constructing the waste management business in the United Kingdom and West Germany, 1945 to the early 1990s", funded by the U.K. Economic and Social Research Council, (ESRC Project Reference RES-062-23-0580).

ACKNOWLEDGMENTS

In the course of several years of research and writing, I have run up a number of debts. Financial support for the project came from the Ohio State University, its department of history and its college of humanities; the Institut für Europäische Geschichte in Mainz; the Deutscher Akademischer Austauschdienst; and the German Historical Institutes in Paris and London. Thanks, too, to my parents for occasional loans to fill in the periods between grants.

My contacts with archivists in several different archives have frequently been rewarding. Special thanks go to Herr Dr. Lenz in the Bundesarchiv, to Herren Peter Göb and Michael Pohlenz at the Bayerwerksarchiv, to Frau Becker in the Historisches Archiv of the Metallgesellschaft A.G., and to Frau Dr. Wolf in the Degussa-Firmenarchiv.

Numerous people read and commented upon all or part of the outlines and/or manuscript. They included Werner Abelshauser, James Bartholomew, Volker Berghahn, Alan Beyerchen, David Blackbourn, Mansel Blackford, Knut Borchardt, William Childs, James Diskant, June Fullmer, John Gillingham, Richard Hamilton, Peter Hayes, Gary Herrigel, Jonathan Liebenau, Brian Linn, Ralph Melville, Alan Milward, Eamonn Noonan, Diethelm Prowe, Claus Scharf, Anne-Marie Stokes, Friedrich Stratmann, Martin Vogt, and Ulrich Wengenroth. Discussions with some of these and other scholars at the Institut für Europäische Geschichte in Mainz, the Ohio State University Department of History, and the Lehrstuhl II für Sozial- and Wirtschaftsgeschichte in Bochum improved the manuscript and helped clarify my ideas. The fact that I did not always take their good advice absolves them of the responsibility for the faults that remain in the final draft.

I would also like to express my thanks to the staff at the University of California Press with whom I have worked. Bettyann Kevles and Diana Feinberg saw the manuscript through the approval process much more quickly than I expected; Shirley Warren and Doug Gower suggested numerous stylistic and substantive improvements to the text. I have thoroughly enjoyed working with all of them.

Finally, I would like to acknowledge special intellectual and personal debts to several persons. I could not imagine a better working relationship than the one I have enjoyed with my adviser, Alan Beyerchen. He pointed me in the direction of this thesis, and then let me find my own way. At the same time, he was available throughout the writing process to discuss problems and to provide feedback on my written work. Without his input, my dissertation would have been entirely different; its relatively rapid transformation into a book would have been impossible. My friends and family have suffered my bad moods and depression from near and far, and supported and encouraged me throughout the research and writing. Finally, I owe the most to Anne-Marie. She has had to live with me through the entire process of writing the dissertation and turning it into a book; her company and good advice have made that process much more enjoyable. Thank you!

Troy, New York
November 30, 1987
R. G. S.

INTRODUCTION

Control of the postwar German economy engaged governmental policymakers during and after World War II. Both the Morgenthau Plan and the Marshall Plan—in diametrically opposed ways—addressed this complex matter, and each of the four Allied occupation authorities developed its own policy to impose effective control. Eventually the key to European peace and prosperity seemed to lie in containing the German economy while at the same time allowing its recovery to serve as the economic motor for all of Europe. By the late 1940s the Germans themselves were also attempting to strike a balance between economic might and political control.

How best to control the German economy was in effect a question of how best to control German industry. The classic example, a crucial step toward achieving the goal of balancing between economic expansion and security, was the creation of the European Coal and Steel Community (ECSC) in the 1950s. The primary object of the ECSC was to harness the enormous economic potential of German heavy industry with strict political controls by Germany's neighbors. Nevertheless, despite the centrality of German industry to the political economy of postwar Europe, we still know too little about it.[1]

Scholars to this point have almost entirely neglected one of the defeated nation's most thriving and creative industrial sectors, the chemical industry. Their lack of attention to its fate in the occupation period and to its postwar reconstruction is surprising. For one thing, the question of what to do with the dominant Germany chemical firm, the I.G. Farbenindustrie A.G., formed a central concern of all of the Allies even before the end of the war. As the largest chemical firm in the world and the largest industrial corporation in Germany, the I.G. and its representatives had been linked closely to the policies and programs of the Nazi regime. Allied authorities knew from the outset of the occupation that successor corporations to the giant concern would play an important role in the postwar development of the German chemical industry regardless of what policies would eventually be implemented. Production from the industry, after all, contributed decisively to German export, and sales abroad would be crucial to German economic survival. The I.G. had been responsible for more than half of total German chemical industry export for most of the prewar period; its successors would have to contribute to postwar Germany's foreign earnings.

Policy on the chemical industry, and on the I.G. in particular, formed a vital component of the management of German economic recovery in the four zones of occupation administered by the British, French, Americans, and Soviets. At the beginning of the occupation, fostering chemical production was necessary to help pay for food imports and to prevent the German zonal economies from collapsing completely. Germany's chemical factories provided fertilizers to replenish spent soil, medicines to stave off disease and cure illness, photographic paper and supplies to help the Allies identify and control the civilian population, and dyes to transform *Wehrmacht* uniforms into attire acceptable to the victors. Later, when Allied policies had changed considerably, western occupation authorities encouraged German representatives from state and industry to think once again in terms of the international as well as the domestic market. Major West German chemical companies returned to prewar business strategies by investing heavily in basic and applied research, concentrating on product quality, and aggressively re-entering foreign markets. They were particularly effective: the industry set the pace for the *Wirtschaftswunder*, or economic miracle, of the 1950s, and continued to outperform most other industry branches in West Germany into the 1970s.[2]

This study addresses three major questions. First, what was the relationship between Allied policy and the reconstruction and resurgence of the West German chemical industry? Second, what factors accounted for the recovery and resurgence of the industry, and what was their relative significance? Third, what light can the examination of the West German chemical industry shed on broader issues associated with postwar European reconstruction? The issues include the development of Franco-German relations, the role of the United States in European affairs, and the impact of the Cold War.

For several reasons, I limit the study to the successor companies

Introduction

to the I.G. Farbenindustrie A.G. in the western zones of occupation from 1945 to about 1951. Since each of the three major successors to the I.G. in western Germany was located in a different zone of occupation, one can study the role of differing western Allied policies in the control and reconstruction of the firms. I do not focus on the I.G. because it is representative: the giant concern is of interest because its production—in terms of both quantity and quality— and its significance to both Allies and Germans dwarfed the remainder of the German chemical industry. Events in the western zones of occupation constitute the primary concern of the study. Records of the I.G. successor factories in the Soviet zone were inaccessible. What is more, Soviet policies toward those factories separated developments in their zone from those in the west. Soviet occupation troops radically transformed the economy in their zone, and, especially in the chemical industry, they dismantled physical plants on a scale unseen in the west. The study's temporal boundaries are 1945 and 1951. While I have been able to establish some lines of continuity from prewar and wartime Germany to the postwar period, the end of the war is a clear point of demarcation. Choosing 1951 as an endpoint was more difficult, although there are good reasons for selecting it: all three of the major successors came into legal existence once again in December 1951/January 1952; the breakup of the I.G. was therefore settled for all practical purposes, and those who would lead the firms through the steady expansion of the 1950s were already in positions of power; and, in the meantime, the industry had already surpassed its prewar levels of output and was well on the way to the resurgence of the 1950s.

I have divided the book into four parts, which in turn comprise eight chapters in all. Part I, consisting of the first chapter, deals with the origins and development of the German chemical industry, the founding of I.G. Farbenindustrie A.G., the firm's relations with the Nazi regime, and I.G.'s postwar planning. Part II, composed of chapters 2, 3, and 4, examines the I.G. successors in the western zones in the initial occupation period. Since differences in occupation policy and practice among the western Allies were most pronounced during the early occupation, I address each zone separately. The arrangement also allows an in-depth introduction to each of the major successor groups to the I.G. and a comparison of the effects of Allied policy on each of them.

Part III, comprising chapters 5, 6, and 7, concerns itself with the I.G. successors from 1947 to 1951, when they returned to world competitiveness. A number of decisive developments characterize the period: the growing impact of the United States on occupation policy; the growing influence of Germans from government and industry on developments in their own country; the continuing coal crisis; and, finally, a series of political and economic turning points (such as the currency reform and the Korean War) that had serious effects on the industry and its chances for recovery. Because these major developments after 1947 had an impact on the successor firms regardless of location, I divide Part III chronologically rather than by zone. Chapter 5 deals with the period from the founding of the combined British and American economic area, the Bizone, to the currency reform in June 1948. Chapter 6 assesses the events of mid-1948 and their ramifications into 1950. Chapter 7 looks at the West German chemical industry at the onset of the Korean War, and at the dimensions and bases for the resurgence of the I.G. successors. Part IV, the final chapter, summarizes the study and explores its wider implications.

PART I
THE BACKGROUND

In January 1933, the National Socialist German Workers Party came to power in Germany. Led by Adolph Hitler, the Nazis sought from the beginning to increase German "living space" by aggressive expansion to the east. Autarky, or economic self-sufficiency, served as both the means to and the end of territorial expansion. Encouraging industry to overcome Germany's poor resource endowment through technological prowess served to supply critical military materials, and new territory would eventually expand the Reich's resource base.

Germany's largest industrial corporation, the chemical firm I.G. Farbenindustrie A.G., was central to Nazi policies. In the years between 1933 and 1939, I.G. Farben's exports provided desperately needed foreign exchange. The firm's technological achievements allowed production of synthetic materials to substitute for commodities scarce in Germany.

The I.G. and its component firms had long traditions in both foreign trade and technological excellence. Founded in the 1860s as part of the "Second Industrial Revolution" that infused scientific research into technology, the firms that would later form I.G. Farben surged to world dominance of organic chemical manufacturing and sales by 1913. They played a pivotal role in the monumental growth of German economic might in the late nineteenth and early twentieth century. One of the component firms, Badische Anilin- und Soda Fabrik (BASF), brought its pathbreaking nitrogen fixation operation to industrial scale by 1915 to help stave off impending German defeat in World War I which would have otherwise been hastened from lack of ammunition and fertilizer. During the 1920s, BASF and then Farben began producing synthetic petroleum from abundant German coal and laid the foundations for production of high-quality synthetic rubber. These impressive technological achievements were part of the reason that National Socialists could dream of German self-sufficiency.

Thus, in order to understand I.G. Farben's place under the Nazi regime, it is necessary to examine in more detail the origins and growth of the German chemical industry and the founding of the firm. On the other hand, Allied policies toward the successors to I.G. after World War II are comprehensible only on light of the corporation's actions in the Nazi era. The history of the chemical industry and of the concern itself influenced policymaking in the postwar period, and therefore deserves a detailed overview.

1
THE RISE AND PRECIPITOUS FALL OF GERMAN *GROSSCHEMIE* 1860-1945

Writing in the aftermath of World War I, Victor Lefebure, a former British chemical liaison officer, labeled the German chemical cartel I.G. Farben "The Riddle of the Rhine." He characterized it as "a serious menace . . . ," "a monster camouflaged floating mine in the troubled sea of world peace."[1] After 1925, when the cartel became the largest German corporation, I.G. Farbenindustrie A.G., it provoked even greater trepidation among some elements of European society. Helmuth Wickel, a left-wing German journalist, warned in 1931 of the immense power and unlimited objectives of the giant concern:

> It is in no way accidental that the I.G. has become the pacesetter for German foreign policy whether in regard to Germany's relationship to France or, even more, in the preparations for creating a European bloc. Leverkusen, Oppau, and Leuna have grown up to become I.G. Germany. They are growing and branching out and are seeking to become I.G. Europe.[2]

More than three decades after the end of the war, the I.G. remained a topic of political passion. Joseph Borkin's best-selling book on *The Crime and Punishment of I.G. Farben* (1978) vilified "I.G.'s embrace of Hitler" and the firm's "moral descent" during the Nazi years. Borkin, who had been a member of the prosecution staff at the trial of the I.G. leaders in Nuremberg, concurred with the opinion of the dissenting judge at the trial that "every member of the I.G. managing board should have been found guilty of mass murder."[3] Writing in the same year as Borkin, a group of West German Marxists characterized I.G. executives as "the acquitted war planners" who had reaped "profit from crises, wars, and concentration camps."[4]

What was it about Farben that called forth such passionate indictments from such varied commentators not only during its existence, but more than three decades thereafter? The answer lies in the firm's history.

THE GERMAN CHEMICAL INDUSTRY, 1860-1925

Through the 1850s, chemical manufacturing in the German states was not particularly highly developed. British, French, and Belgian entrepreneurs were the first to construct a large-scale and diversified chemical industry.[5] Even in the first half of the nineteenth century, however, strong indications of the German area's potential existed: it was blessed with rich coal reserves, especially in the Ruhr region, and other raw materials (e.g., pyrites), labor, and capital were available as well. Lack of many important raw materials provided a stimulus for innovation for the industry. By the 1840s, chemists were available to pursue such innovation. Practically oriented chemical education, developed under Professor Justus von Liebig, contributed to the "large reservoir of keenly enthusiastic chemists" so necessary for the subsequent growth of chemical manufacturing.[6]

During the 1860s, the organic chemical industry in the German area began its spectacular expansion. Industrialists combined small amounts of foreign raw materials with native coal, water, air, and ingenuity and parlayed the mixture into new products. Most of the firms that would later form the I.G. Farbenindustrie A.G. were founded at that time. Friedrich Bayer & Co., Meister Lucius & Brüning (later Hoechst A.G.), and Kalle & Co. all came into existence in 1863. Friedrich Engelhorn founded the Badische Anilin- und Soda-Fabrik (BASF) in 1865, and the Farbenfabriken ter Meer at Uerdingen dates from 1877. In 1887, 12 chemical firms figured among the 100 largest German corporations. They all came into existence between 1858 and 1881. Moreover, they tended to stay on the top: in 1907, all 12 were among the top 100 German firms. Altogether 14 of the 17 chemical firms in the top 100 in that year had been founded in the same time frame (1858-1881).[7]

The genesis of so many successful firms within such a brief time span resulted from the confluence of several factors. The Germans were lucky enough to remain untroubled by the patent litigations that plagued the French and English chemical industries in the 1860s. Furthermore, a combination of the luck of the newcomer and the skill of the developer and salesman meant that the Germans were the first to produce synthetic alizarin dyes, used primarily by French and British textile manufacturers on a large scale. German universities had, in addition, trained a large cadre of chemists for use in the industry. Finally, German entrepreneurs commanded astonishing acumen in all areas of finance, research, production, and sales.[8]

The birth of the "modern German chemical industry" in the 1860s involved the exploitation of an entirely new area of manufacturing, organic chemicals. Based ultimately upon coal, organic chemicals were generally made from coal tars, benzene and other volatile materials, and coke, all of which were the products of coking and carbonization operations.

As a result of these common raw materials—and owing as well to basic similarities in hardware, such as reaction vessels and transfer equipment—production of organic chemicals featured extensive interconnections among various production technologies. A closer look at the products of benzene (utilizing processes developed to 1945) clarifies this. Crude benzene, for instance, could be treated to form chlorobenzene, which was the basis for production of some plant protection agents. Chlorobenzene itself served as an intermediate in manufacture of phenols, which in turn constituted the basis of salicylic acid, tanning agents, and various synthetic products. The same intermediate was used for production of nitrochlorobenzene, which was indispensable for manufacturing pharmaceuticals and photographic chemicals. Benzene intermediates also served as preliminaries for dyestuff, solvent, and pharmaceutical production.[9]

After the founding of key firms in the 1860s, it was one thing for German chemical industrialists to start the rapid expansion and another to sustain it. Later in the century, Germany had to take a number of measures to ensure continued growth. The newly unified German state aided the industry by maintaining a reasonable financial policy, prudent yet generous enough to encourage capital formation. It also permitted—and even actively encouraged—

agreements to restrict destructive competition. In addition, the state provided the chemical industry with a workable patent law. The Reich Patent Law of 1877 "specified both a patent search, to ensure novelty and substance in patents, and the limitation of protection to processes, as opposed to products." Process rather than product protection afforded chemical industrialists an incentive to improve upon existing processes and "also specified substantial and increasing fees, to ensure that a patent holder did not block others while failing to work the patented process himself. Finally, added protection was provided the patent holder by *Zusatzpatente* after 1891, covering applications and improvements."[10]

These explanations for the enormous expansion of the industry to 1900 need amplification, however. Easing of rivalry in the domestic market through cartel agreements, for instance, did not eliminate destructive competition. German firms still competed ruthlessly among themselves at home and especially abroad: the Reich encouraged cartelization domestically in order to allow German firms to compete in crucial foreign markets. Moreover, in the creation and expansion of their chemical industry, the Germans were fortunate enough to enter the international competitive arena at a late enough date to allow them to learn from the mistakes of others and to benefit from the most recent developments within the industry. At the same time, they arrived on the scene *early enough* so that the industry and its firms were still small. Leading chemical firms the world over were not yet well established or beyond the reach of German capitalists.[11]

Expansion of the chemical industry in the German area was impressive. Between 1872, the year after the founding of the German Reich, and 1913, the year before the outbreak of World War I, production in the industry increased on average by 6.2 percent per year. In contrast, the yearly rate of increase for all German industrial and artisanal production between 1870 and 1913 was just 3.7 percent on average.[12] Since high profitability accompanied relatively large growth rates, the organic chemical industry, unlike other industrial sectors, was able to finance much of its own expansion. Companies paid out enormous dividends to stockholders—on the average, about 20 percent of their stock capital at the turn of the century. They used stock capital and retained earnings to avoid dependence upon and control by banks. Through 1945, relative independence from banking capital characterized the German chemical industry.[13]

Growth in German chemical production was not without its side effects. As the industry surged to a position of world leadership in dyes and many other important production areas during the 1890s and the early part of this century, other countries began to react with hostility to the German advance. In many cases—for example, in France and Russia in the late nineteenth century—tariff barriers appeared. Given their limited internal market, German firms, dependent upon export, responded to protectionism by erecting foreign production facilities. They much preferred to sell their products through subsidiaries, but built factories abroad as a defensive reaction. Limited foreign production facilities make it difficult to characterize the industry as multinational before 1945.[14]

German chemical manufacturers competed with one another even though they could conclude legally binding contracts to limit competition. All of the firms competed for foreign sales. To improve their sales positions near the expiration date of cartel agreements, they began to quarrel among themselves and to undercut one another. Prices of chemical manufactures dropped steadily to the point where less efficient German chemical companies fell into acute financial difficulties; larger, more cost-competitive firms looked for a way out of the dilemma. Carl Duisberg, one of the first to recognize the problem, proposed a solution.

Born in 1861, Duisberg came to work for the Farbenfabriken Bayer at Elberfeld as a research chemist in 1884.[15] By the early 1890s, he made a name for himself through his inventions and through his memorandum on the development of the firm's newly purchased site at Leverkusen. Duisberg became a member of Bayer's managing board at about the same time and, following a trip to the United States in 1903, produced his pathbreaking 1904 memorandum on the future of the German chemical industry. Impressed by John D. Rockefeller's Standard Oil Trust in the United States, he suggested that the major firms of the German chemical industry merge in order to compete more effectively on world markets and to avoid ruinous competition at home. Almost immediately, movements in this direction began with the formation of the *Dreibund*, a loose confederation of three major chemical companies, Bayer, BASF, and Agfa. The new organization fell short of Duisberg's vision: component firms remained privately controlled; production planning, marketing, and pricing stayed in the hands of the individual companies. Many of the most important firms in Germany stayed aloof from the *Dreibund*. In fact, it called into being a rival association, the *Dreiverband*, formed between the firms of Hoechst, Casella, and Kalle between 1904 and 1907. Twenty more years would pass before Duisberg's dream was fully realized. In the meantime, the industry continued to grow and, owing to unforeseen developments, to change in nature.

On the eve of World War I, the German chemical industry had conquered several key world markets. Chemical exports from the Reich constituted about 28 percent of total world chemical exports and represented between 10 and 15 percent of Germany's total exports. German firms virtually monopolized world trade in dyestuffs and other organic chemicals and were extremely important in world pharmaceutical sales. They also stood at the forefront of new developments in the rapidly changing industry, pioneering in the development of synthetic nitrogen production and in other areas of high-pressure chemistry.[16]

The war itself significantly altered the structure of the industry. Wartime demands necessitated pronounced expansion into important new production areas. If World War II (owing to the development of the atomic bomb and radar) was the "Physicists' War," World War I was the "Chemists' War." Especially in Germany, the chemical industry drew on its tradition of helping transcend the country's relatively poor resource endowment. It supplied various synthetic products, including artificial butter, honey, and rudimentary plastics. Production of synthetic ammonia proved more important for the war effort itself. The story of the rapid translation of this new production technology from pilot plant to full-scale production has been told many times,[17] but it is worth emphasizing that the rapid implementation of the synthetic nitrogen production program under the leadership of BASF's Carl Bosch created German self-sufficiency in this essential product (crucial to the production of fertilizers and ammunition) and helped ensure German survival during more than four years of total war. In the absence of the program, German supplies of nitrogen would probably not have lasted through 1915. Running out of nitrogen would have guaranteed early German defeat.

The industry contributed another key product to the war effort. On April 22, 1915, Allied troops on the western front near Ypres in Belgium received the first surprise dose of a new weapon from the chemist's arsenal, poison gas. The incident was the first salvo in an escalating technological battle between the Central Powers and the Allies. Each side developed new gases and, at the same time, new gas masks, the major defense against gas attacks. Thus, while the number of casualties caused by gas attacks spiraled upward, the percentage of deaths decreased. The Germans, relying chiefly on technological prowess and production experience drawn from their dyestuffs industry, developed poison-gas production technology that was superior to that of the Allies, although the two sides were almost equal in gas protection techniques.[18] Germany's potent poison gas arsenal affected the final outcome of the war and the history of the Weimar Republic. A British technician estimated that 50 percent of German shells in 1918 contained gas, and their use probably allowed German soldiers to retreat in good order. That orderly retreat gave rise to the *Dolchstoss*, or stab-in-the-back, legend, to which many elements of German society subscribed in the 1920s.[19]

Increased competition from abroad after 1918—another result of the war—altered the structure of the international chemical industry. Before the conflict, German firms supplied many crucial basic and intermediate chemical products to other countries for further processing. Beginning in 1914, the British shipping blockade forced Germany's customers—even those who remained neutral—to develop their own industrial base for producing chemicals. In the United States, dangerous competitors to German industry emerged, including the vastly enlarged and diversified Du Pont corporation.[20]

The Great War had a further structural effect by continuing and intensifying the movement toward concentration within the German chemical industry. Carl Duisberg recognized its long-term ramifications and, as he had before, proposed a course of action for the industry. In an update of his 1904 memo written in August 1915, Duisberg stressed that the war would pose serious challenges after the hostilities came to an end, anticipating, among other things, ruinous competition from abroad. What is more, though, the war effort had involved the extensive expansion of capacity in certain areas. Capacity to produce synthetic ammonia, for instance, would in all probability outstrip postwar demand.[21]

This set of difficulties would arise from the war itself. Duisberg also emphasized long-term trends within the industry, however, some of which the hostilities had exacerbated. In 1915, the German chemical industry appeared financially and technically sound. Large profits in the prewar and initial war years, however, had come primarily from sales of dyes. Patents on dye production processes would soon expire.[22] At the same time, costs within the industry were on the rise: labor and research expenditures were high and mounting; providing good service to customers made large—and expensive—inventories of chemical products unavoidable; foreign subsidiaries were necessary to protect market shares but also involved significant costs. Since each of the major companies had subsidiaries in the same countries, effort was frequently duplicated.[23]

Duisberg's analysis pointed to the same solution to the problems of 1915 as that proposed in his earlier memo: fusion of all major German chemical firms into a single concern. He admitted that the strategy had its shortcomings, including the possibility that the public and the state would come to hate and fear the giant firm. What is more, absence of competition within the German chemical industry could lead to laziness and lack of creativity.[24] Careful attention to the structure of the company (for instance by encouraging competition within the firm) would, however, minimize the negative consequences of the strategy.[25] In any case, as he saw it, the advantages of fusion vastly outstripped the drawbacks. Common sales and purchasing organizations, the absence of interfirm patent litigation, coordinated and cooperative research and development efforts, and unification of foreign subsidiaries would be the most important payoffs of the legal fusion of major chemical firms.[26] Duisberg emphasized repeatedly throughout the memo that complete fusion would have the effect of lowering costs, setting aside competition, and ensuring high profits for all firms concerned.[27]

Pressures arising from the conduct of the war brought about more extensive cooperation among the industry's largest firms along the lines Duisberg proposed. In 1916, the two groups formed earlier by major German chemical firms, the *Dreibund* and the *Dreiverband*, combined to form the *Interessengemeinschaft der deutschen*

Teerfarbenindustrie (Community of Interest of the German Coal Tar Dye Industry), which later came to be known as the "Little I.G." The new organization proved rather less satisfactory than what Duisberg had hoped for since the component companies remained separately owned and operated, resulting in decentralized and often uncoordinated decision making. But the firms divided up production quotas and profits and shared research developments and technology.

Only after the armistice in 1918 did the industry begin to feel the full impact of loss of market share abroad, increased foreign competition, and overcapacity. The defeat of the Central Powers had still wider effects. The Treaty of Versailles included provisions restricting German control over foreign economic policy until 1924 and thus posed yet another obstacle to the recapture of foreign markets for chemical firms. What is more, the Germans had financed the war effort on the assumption that they would be victorious and would therefore be able to force their defeated enemies to take over part of the costs of the conflict. Government spending during the war therefore vastly exceeded income from tariffs and taxes, and Germany's ability to pay back debts incurred during the hostilities waned. The resulting inflation, held in check to some degree until 1918 by wartime controls and appeals to patriotism, ran rampant in the postwar year. By 1923, inflation had reached disastrous dimensions and rendered conduct of business within Germany extremely difficult.[28] Monetary reform brought about more stable business conditions by 1924, but the years of uncertainty had a disturbing psychological effect on business. Currency reform entailed, in addition, problems of liquidity for most companies.

To combat all of these problems, the firms of the Little I.G. finally decided in 1925 to implement the fusion plan that Duisberg proposed in 1904 and reiterated in 1915. Each of the major component firms sold its holdings to the Badische Anilin- und Soda-Fabrik in Ludwigshafen; the new corporation, renamed I.G. Farbenindustrie A.G., transferred its seat to Frankfurt am Main (see fig. 1). It began operations on December 9, 1925.

THE I.G. FARBENINDUSTRIE A.G.

Although unprecedented in terms of size and power, the newly consolidated I.G. Farbenindustrie continued in the initial years of its

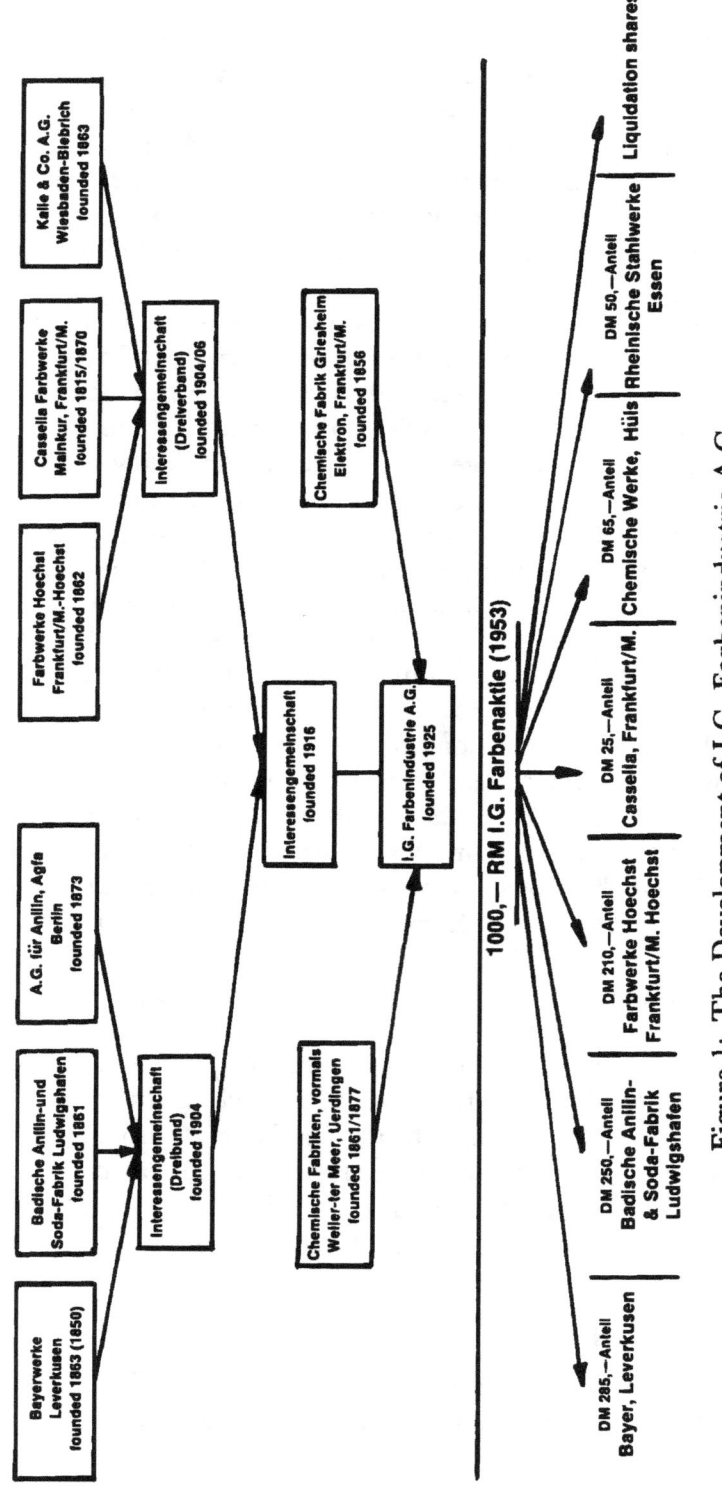

Figure 1: The Development of I.G. Farbenindustrie A.G.

Source: W. O. Reichelt, *Das Erbe der I.G. Farben* (Düsseldorf: Econ, 1956), p. 100.

existence many of the traditions of German chemical manufacturing. The combine relied heavily on research to develop new product lines, and it spent accordingly: through 1930, *after* the onset of the Depression, Farben's research expenditures amounted to between 8 and 12.7 percent of its turnover and frequently exceeded net profits.[29] High profitability was another hallmark of the German chemical industry that the I.G. upheld. Throughout its existence, net profits never sank below 5.4 percent of turnover and, in the years after 1935, regularly exceeded 10 percent of sales volume.[30] The firm was thus able to maintain its long-standing independence from the influence of banks. Finally, as an export-oriented firm in a perennially export-driven industry, the I.G. generally supported doctrines of economic liberalism. Carl Duisberg and Carl Bosch, the two most prominent representatives of the firm until the mid-1930s, advocated free trade and limited state intervention in the economy.[31] At the same time, neither the industry nor the firm was averse to accepting government subsidies for projects (such as synthetic nitrogen) that permitted Germany temporarily to overcome its poor resource base.

By virtue of sheer concentration of economic might, however, I.G. Farben represented a break with the German past. It constituted "by far the largest German business organization . . . [and] belonged . . . among the group of the largest firms that the world knows."[32] The giant American concerns, U.S. Steel, Standard Oil of New Jersey, and General Motors, dominated this select group and vastly outstripped the I.G. in size in terms of working capital and turnover. But Farben was third in number of employees and was perhaps the largest firm in the world in its turnover from exports (comparable figures for Standard Oil's turnover from exports are not available) from its founding until 1934, when GM took the lead. In terms of turnover, export turnover, and employees, the I.G. was the largest *chemical* corporation in the world from 1926 to 1944 and not far behind rivals I.C.I. and Du Pont in terms of working capital.[33]

The new company dominated the German chemical industry. From 1926 to 1938, it possessed between 45 and 48.5 percent of the total capital invested in all German chemical firms and produced between 25.3 and 31.9 percent of the chemical industry's total turnover. Its share of the lucrative export market was even more impressive. Sales abroad rose from 5.5 to 8.7 percent of total German exports from 1926 to 1938, and generally amounted to well over half of the export turnover of the entire German chemical industry: in 1938, its international sales accounted for 57.5 percent of total German chemical industry export turnover. The I.G. employed 114,000 persons in 1928, while the next largest chemical producer, Deutsche Solvay-Werke, employed just 6300. In 1928, the giant concern's assets were valued at RM 1753 million, those of Deutsche Solvay-Werke at RM 134 million.[34]

Farben's dominance of the German chemical industry did not extend to all fields: even though the firm produced a wide variety of chemicals comprising twenty different major production areas and ranging from dyes to light metals,[35] its share in a given area was not uniform. During World War II, it manufactured almost 100 percent of Germany's dye and synthetic rubber needs, around 70 percent of its nitrogen, 60 to 70 percent of its photographic supplies, and about half of its pharmaceuticals. It also controlled about 90 percent of the country's organic intermediates production and (together with its licensees) about 90 percent of its synthetic gasoline manufacture.[36] I.G. left the bulk of production of some inorganic chemicals (e.g. sulphuric acid, chlorine, and potash), of soap, and of paints to smaller German chemical companies.[37] In general, Farben's management concentrated production and investment in areas of high profit, most often on the cutting edge of new technologies. Other firms manufactured less profitable and less sophisticated items, or worked their way into certain specialty niches (e.g., Merck and Böhringer in pharmaceuticals).

Production in such diverse areas forced the I.G. to become acquainted with diverse markets, ranging from sales to other producers (e.g., dyes and organic intermediates) to direct consumer sales (e.g., pharmaceuticals and photographic supplies). Market complexity therefore mirrored interwoven production relationships and sophisticated technology. Sales in varied foreign markets through I.G.'s numerous subsidiaries[38] further magnified the intricacy of the firm. Its market, product, and technical complexity marked its organization.

Figure 2 shows a streamlined version of the company's organization.[39] At the top of the firm stood the standard, legally required *Aufsichtsrat* (supervisory board) and *Vorstand* (managing

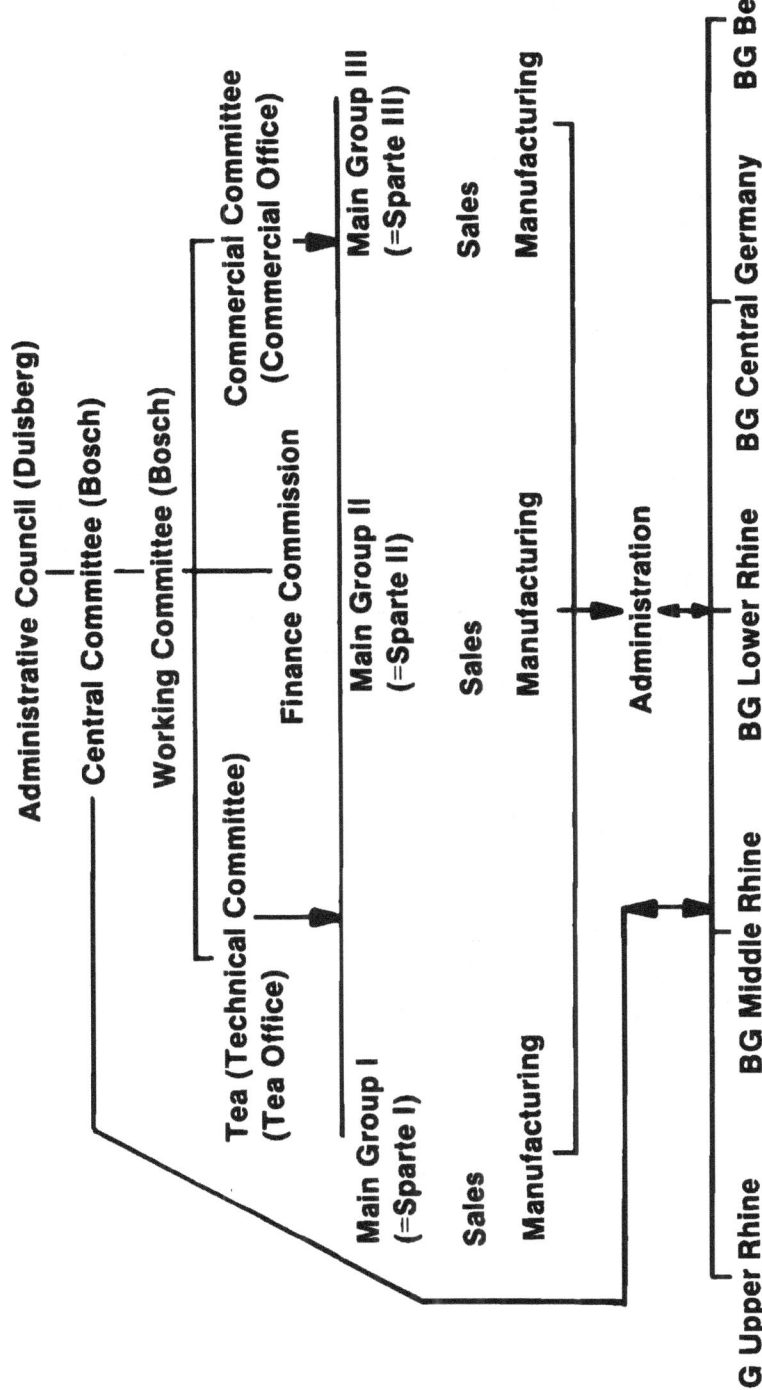

Figure 2: The Organization of the I.G., September 1931.

[a] BG = Betriebsgemeinschaft, or Works Group.

Source: Adapted from Helmuth Tammen, *Die I.G. Farbenindustrie Aktiengesellschaft (1925–1933)* (Berlin: Tammen, 1978), facing p. 28.

board). In the I.G., both differed somewhat from the norm for German joint-stock corporations, for they exercised surprisingly limited power. The supervisory board consisted of about fifty members, an unwieldy number. As a consequence, a *Verwaltungsrat* (administrative council) composed of just eleven board members took over most of the supervisory functions. In practice, even the council exercised little authority: only the chairman had real input into policymaking. Until 1934, Carl Duisberg occupied this position, Carl Bosch from 1934 to 1940, and Carl Krauch thereafter.[40]

Even more cumbersome than the supervisory board was I.G.'s managing board with eighty-three "members and deputy members" at the beginning of the concern's existence in 1925. It had in any case no real opportunity to exercise any authority, since, until 1937, it never met.[41] Instead, the *Arbeitsausschuss* (working committee), headed by the chairman of the managing board (and later the *Zentralausschuss* [central committee]), held ultimate responsibility for policy. The working committee made many of its decisions on the basis of recommendations from two central committees, the Technical Committee (TEA), composed of the technical members of the managing board, and the Commercial Committee (KA), composed of the board's commercial members. Several agencies supplemented the core organization, including those responsible for legal affairs, insurance, transport, purchasing, accounting, advertising, and press relations. Central offices also existed to keep watch over developments in the economy (VOWI) and in politics (WIPO).[42]

The I.G.'s central structure exercised some degree of control over sales and production. Products were divided into three main groups, or *Sparte*, depending upon technology employed or production intermediates required. Sparte I controlled sales and set production goals in the field of high-pressure chemistry. Sparte II performed the same function for dyes, pharmaceuticals, plastics, and chemical intermediates. And Sparte III was responsible among other things for photo products and artificial silk.[43]

Despite the centralization of many functions within the firm, extensive control over actual production and over many other key functions remained in the hands of the production groups. The decentralization of the firm into *Betriebsgemeinschaften* (BG), or works groups, is of primary interest here. Initially, four works groups divided up control of the thirty-three major production complexes[44] of the I.G. The groups included the BG Upper Rhine, with its center at Ludwigshafen and composed mainly of former BASF holdings; the BG Middle Rhine (later renamed BG Maingau), headquartered in Frankfurt-Hoechst and for the most part composed of the former Meister Lucius & Brüning corporation; and the BG Lower Rhine, centered at Leverkusen and including the factories of the former Bayer company. Two subgroups—in Wolfen-Bitterfeld and in Frankfurt—together constituted the fourth works group, the BG Central Germany. On November 22, 1929, I.G. management reorganized this scheme somewhat. The Hoechst group obtained the Frankfurt facilities of the BG Central Germany. At the same time, a fifth group, the BG Berlin, was created in Germany's capital to assume control of the BG Central Germany's photo, synthetic silk, and plastics factories.[45]

Each of the directors of the works groups acted as an active member of the managing board and operated independently to a remarkable degree. In fact, as Helmuth Tammen noted in his study of the I.G.,

> the individual I.G. plants [Betriebe] were not only production facilities..., but also constituted the actual centers of every activity in the chemical and technical area. Most factories had laboratories for experimentation and for application of technology, patent offices, legal sections for contracts and local legal questions, personnel and social sections; they conducted their own bookkeeping in the determination of costs and tax write-offs; and they dealt with the purchase of building materials, machines, and apparatus to a certain extent on their own responsibility.[46]

The works groups differed from one another in many ways. For one thing, as indicated by the names of the groups, each tended to cluster in a single region. BG Lower Rhine, BG Maingau, and BG Cental Germany (after 1929) conformed most to the ideal of regional concentration. But there were several anomalies: BG Upper Rhine included a factory in Bochum (actually in the Lower Rhine district) and the enormous Ammoniakwerke Merseburg at Leuna in central Germany. Factories from the Berlin group were scattered throughout Germany, and included plants in Munich and Bobingen.

The groups specialized to some degree according to product

area, although almost all of the central factories of the individual production groups produced a broad range of basic chemicals, intermediates, and finished products. The Ludwigshafen group controlled much of the I.G.'s high-pressure chemistry production, for instance, and the BG Lower Rhine excelled in sales and production of pharmaceuticals. Berlin directed manufacture of much of the firm's photographic paper, film, and artificial silk.

Limited product specialization among the works groups hinted at another, somewhat more nebulous dimension of differentiation: by tradition. During the I.G. period, the relatively new Berlin and central German groups developed their own standard operating procedures. The remaining three groups centered around firms with long-favored manufacturing practices and product areas and dominant styles of management. Both the Ludwigshafen and Leverkusen groups had lengthy backgrounds in certain product areas, as noted above, and gained prominence also for their research activities. BG Lower Rhine also had a well-deserved reputation in the area of foreign trade. In addition, leadership of each of the three groups tended to come from within. Complete biographical data on all of the management personnel of the three is virtually unobtainable. Nevertheless, a clear career pattern emerges from available information. Each of the wartime directors of the western works groups spent his entire career within the predecessor firm to the works group (or within the works group itself). Each postwar director of the three major successors to the I.G. began his career within the works group in question. Similar patterns hold true for many members of the managing boards of the successors.[47]

The trust's central organization—through control of sales and export agencies, foreign and domestic subsidiaries, financial institutions, research facilities, and information—therefore served to develop strategy, to coordinate production, and to oversee financial affairs for an agglomeration of relatively independent production units. Each of the component groups had its own management staff and style of manufacturing. The directors of each were prominent members of the I.G. managing board and controlled much of their own purchasing, production, and even research. In a very important sense, then, the I.G. was never a fully unified concern.

Technical interrelationships among product lines in a given group,[48] firm tradition and leadership, and the proximity of a given

Table 1. Investment in New Plant by the I.G., 1933–1944
(in million RM)

Plant	Investment		
	1933/1936	1937/1940	1941/1944
Schkopau	10	212	179
Heydebreck	—	13	343
Auschwitz	—	—	322
Hüls	—	127	133
Moosierbaum	—	2	102
Ludwigshafen (Buna III)	—	—	91
Landsberg	—	12	60
Stassfurt	16	20	14
Aken	23	9	14

Source: "Investment for New Plant of I.G. from 1933–1944 in Million RM." (Adapted from bar chart.) August 18, 1945. Public Record Office, London (Kew), FO 1031/233.

factory to transportation and raw materials—in other words, a host of commercial, economic, and personal factors—had a pronounced effect on the development of the individual production groups. Tables 1 and 2, which present figures on investment that are reasonable in terms of order of magnitude even though not based on official I.G. documents, demonstrate this clearly.[49] Between 1933 and 1944, factories located in central Germany received the lion's share (over RM 1.35 billion total) of investments in new plants in the I.G. (see table 1). Their bounty rested upon several factors: proximity of coal; distance from potential air attacks from the west; desire to make effective use of technological interdependence by developing large-scale production of related processes; and, in the case of Auschwitz in particular, pressure from the regime and the presence of plentiful and cheap slave labor.[50] In the west, only the Hüls plant, which belonged to the BG Upper Rhine by virtue of its high-pressure technology, obtained such extensive investment funds. Again, proximity of coal supplies for fuel and raw materials and desire to make use of technological interdependence played a role. Of the established factories in western Germany,

Table 2. Investment in Large Plant by the I.G., 1925–1944 (in Million RM)

Zone	Plant	1925/1929	1930/1932	1933/1936	1937/1940	1941/1944
United States	Hoechst	57	16	35	59	53
French	Ludwigshafen	71	8	48	76	124
	Oppau	48	2	17	76	91
	Lu-Buna III	—	—	—	—	91
Great Britain	Hüls	—	—	—	127	133
	Leverkusen (incl. Titan)	54	9	31	58	45
	Uerdingen	6	2	7	16	21
	Augusta Viktoria	10	4	8	25	19
	Dormagen	33	2	6	11	15
	Duisburg	10	2	9	22	8
	Knapsack	8	1	9	18	2
	Gruppe I					
	Merseburg	256	26	89	177	229
	Heydebreck	—	—	—	13	343
	Gruben	53	4	58	125	105
	Gruppe II					
USSR	Schkopau	—	—	10	212	179
	Bitterfeld	45	8	42	77	107
	Wolfen Farben	37	2	18	49	34
	Gruppe III					

Wolfen Film	43	7	43	82	22
Piesteritz	26	1	4	3	2
Stassfurt	—	—	16	20	14
Aken	—	—	23	9	14
Landsberg	—	—	—	12	60
Premnitz	15	3	11	16	12
Outside frontiers of 1937 Auschwitz	—	—	—	—	322
Moosierbaum	—	—	—	2	102

Source: "Investment for Large Plant of the I.G. in Million RM." (Adapted from bar chart.) August 18, 1945. Public Record Office, London (Kew), FO 1031/233.

only the Buna III plant at Ludwigshafen received substantial attention from I.G. investment officers to expand its facilities and to make better use of existing production relationships.

A more complicated picture emerges from examination of table 2. Clearly, the large central German plants received heavy investment, especially at Heydebreck, Schkopau, Bitterfeld, and Wolfen Farben. BASF factories also fared well: Ludwigshafen and Oppau (including the Ludwigshafen Buna III plant) attained a total of RM 652 million investment between 1925–1929 and 1941–1944; the group's plant at Hüls obtained RM 260 million in investment from 1937 to 1944; Auschwitz, also under BASF control, received RM 322 million in the period 1941–1944; and, BASF's factory in central Germany at Leuna obtained more financial attention than the group's original factories at Ludwigshafen and Oppau combined—altogether RM 777 million from the fusion until 1944. Initial investment in the Leuna plant in 1925–1929 reflected Bosch's plans to develop large-scale synthetic gasoline production facilities, while cutbacks in the period 1930–1936 reflected the world economic downturn and the fall in world gasoline prices. Later investments resulted from the Nazi regime's encouragement and from the need for various synthetic products turned out by the factory.[51]

The Bayer group participated in the financial bonanza as well, albeit more modestly. Between 1925 and 1945, the main plant expanded and updated to the tune of RM 197 million, while the group as a whole received RM 316 million. The Hoechst group, in contrast to all of the other groups (excluding Berlin, which obtained large-scale investments—RM 197 million—only at its Wolfen Film plant), was relatively neglected, obtaining just RM 258 million in investment funds for large-scale plants from 1925 to 1944. The main plant at Hoechst, of course, fared best within the group, obtaining RM 220 million in investment capital. In other words, the main Hoechst plant received more capital investment than did the main Bayer plant; the Hoechst group as a whole received less than the Bayer group as a whole.

By 1945, the I.G.'s investment strategy for large plants itself accounted for significant differences among the works groups. Although heavy bombing caused extensive destruction to plants belonging to the BASF group, what remained was relatively new technologically and often—since there were heavy investments late in the war—relatively lightly affected by heavy usage in fulfilling German needs. Bayer plants, though less heavily damaged by bombing, were also not as new. Plants in the Hoechst group were often outdated and worn out, if only lightly bombed. The central and eastern German factories had been bombarded and were often severely damaged in the final fighting of the war, but were often the most up-to-date technologically.

THE I.G. IN THE NAZI PERIOD

Although I.G. Farbenindustrie A.G. existed for only twenty years, numerous controversial issues surround the company's history. By far the most widely discussed, the relationship between the firm and the Nazi regime, was largely responsible both for the depth of feeling toward the I.G. noted at the beginning of this chapter and for the postwar treatment of the corporation at the hands of the Allies. In reality, however, the relationship between the I.G. and Nazis involves two separate questions: Did personnel from the I.G. aid the Nazis in coming to power? and, Did Farben officials lead or follow in Nazi-era developments, specifically in the areas of autarky, planning and carrying out an aggressive war of expansion, plundering the European chemical industry, and use of slave labor?

Representatives of the firm did not directly help the Nazis come to power in 1933, although circumstantial evidence exists to support the contention.[52] In the late 1920s and early 1930s, the corporation faced hard times owing to the Great Depression. Extensive financial commitment to synthetic fuels development worsened matters. Such investment had seemed wise in the early and mid-1920s when expert predictions held that the world would soon run out of oil. But the predictions were wrong: large reserves of petroleum, discovered in Texas and Oklahoma in the mid-1920s, combined with the Depression to drive world oil prices down.[53] For this reason, the I.G. prevailed upon late-Weimar governments to increase import duties on imported oil. Sometime around June 1932, after it became obvious that the Nazis would emerge as serious contenders for power in Germany, Carl Bosch, the chairman of the managing board, decided to approach the party and sent two of his lieutenants, Heinrich Gattineau and Heinrich Bütefisch, to discuss the matter with Hitler. Allegedly a deal resulted from the meeting: the Nazis would, when in power, raise duties on oil to protect the I.G.'s program. In return, the firm would contribute to the Nazi war chest to finance political campaigns.[54] Supposed contributions to the party are not the only grounds on which critics of the firm argue for its complicity in the Nazi seizure of power: they claim, in addition, that the company had among its ranks several persons with close connections to Nazi party members in the years before 1933.[55]

Closer examination of the firm's role in the rise of National Socialism yields a different picture. It is true, for instance, that Gattineau and Bütefisch met with Hitler in 1932 to discuss the Führer's policy on synthetic fuel. They tried, though, only to obtain Hitler's assurance that the attacks of the Nazi press and orators on the I.G., and especially its synthetic fuels program, would stop. Hitler agreed to their request, although verbal and written attacks on the chemical firm by Nazis continued (albeit combined with sympathy toward synthetic fuels development). Apparently no one mentioned alterations in customs duties in the event of a Nazi government; nor did I.G. representatives pledge financial support to the NSDAP.[56]

Small contributions to Nazi political campaigns indeed flowed into party coffers. Max Ilgner of the notorious I.G. NW7 in Berlin (which was the center of the firm's intelligence, propaganda, and political economic operations) gave some of his discretionary funds to the NSDAP through Walther Funk and others, as did other younger and lesser-ranking members of the I.G. middle management. Small amounts trickled into the party treasury through other, indirect channels. Donations were, however, small and indirect in every case; the recipients were probably unaware of where the money came from. What is more, the concern's uppermost leadership was unaware of the contributions. Finally, although the firm was later sprinkled with Nazi members in upper and middle management, only one member of the managing board, Wilhelm Rudolph Mann, belonged to the party prior to January 1933. He kept the fact to himself until after the seizure of power.[57]

All Farben contributions to the Nazis—direct and indirect— dried up in late 1932 as the party took a more radical turn and seemed to lose its appeal at the polls. I.G.'s leaders continued to be apprehensive in general about the party's plans for autarky because of the company's dependence on export.⁵⁸ Obviously, forces existed within the firm favoring accommodation with or even support of the NSDAP prior to its coming to power, but these were lower-ranking and younger members of the leadership group. The firm's upper management tried, however, to keep itself on the good side of the party in the event of Nazi participation in a national government.

The question of who led or followed in the development and implementation of political and economic aims during the Third Reich is a thornier one. U.S. government personnel had convinced themselves that they possessed the answer as World War II came to a close in 1945: the "I.G. Farbenindustrie A.G. played a prominent part in building up and maintaining the German war machine."⁵⁹ Later, in 1947, U.S. prosecutors moved against twenty-three high-ranking officials of the chemical combine, charging them with numerous crimes, including conspiracy to plan and carry out an aggressive war of expansion, spoliation, and crimes against humanity. In fact, however, the "primacy of politics" prevailed in Nazi Germany,⁶⁰ and the I.G.'s leadership was not exempt from party dictates on broad policy objectives (such as if and when to make war). But their relative lack of influence in setting broad political aims did not prevent Farben's leaders from working to attain them. In the process, the firm profited enormously and brought shame upon itself.

Firms exist to make money. In the Nazi period, I.G. Farben fulfilled this function thoroughly.⁶¹ Table 3 demonstrates that the I.G. was an extremely profitable corporation. Net earnings amounted to 11.6 percent of sales turnover in the first full year of its existence. They dropped noticeably during the 1920s and early 1930s. At the beginning of the drop-off, until 1931, distributed profits remained high while the firm ran down its reserves and cut its research budget. Farben slashed its outlays in all areas as the Depression continued through 1932.

The firm's experience during the Depression transformed it fundamentally. After 1930, it never again devoted more than 7 percent of its sales turnover to research. Indeed, throughout the Nazi period, the trend was for research expenditures as a percent of sales to decrease: beginning in 1940, research amounted to under 5 percent of turnover; after 1941, it was less than 4 percent.⁶² The

Table 3. Net Profits of I.G. Farbenindustrie A.G. 1926–1944 (not adjusted for inflation)

Year	Net Profit[a] (million RM)	Net Profit as Percent of Turnover
1926	119.8	11.6
1927	113.6	9.0
1928	110.9	7.8
1929	100.9	7.1
1930	90.2	7.6
1931	55.0	5.4
1932	48.6	5.5
1933	62.9	7.0
1934	64.9	6.6
1935	66.8	6.1
1936	132.3	10.2
1937	195.0	12.9
1938	190.4	11.6
1939	239.2	12.0
1940	298.1	13.8
1941	311.5	12.3
1942	271.3	9.3
1943	300.4	9.6
1944	145.4	5.7

[a] Net profit is defined here as distributed profits (*Dividende*) plus retained earnings (*Reservenstellung*). Taxes, research monies, and write-offs are not included.
Source: Compiled (and percentages calculated) from Hermann Gross, *Material zur Aufteilung der I.G. Farbenindustrie Aktiengesellschaft* (Kiel: Institut für Weltwirtschaft, 1950), Tabelle la.

strategy affected the industry for years to come, since research budgets did not rebound until well into the 1950s. I.G. successors had to work on applications of and improvements on basic discoveries from the 1920s and earlier until their own reserves recovered sufficiently to permit further basic research. What is more, the Depression inaugurated a new era for shareholders of the firm. Although dividends remained fairly high (as a percentage of the nominal value of the stock), they failed to keep pace with burgeoning sales turnover during the Nazi period. Distributed earnings rose only in the final throes of the war (1942–1944).[63]

Nazi economic policies—which continued in many respects the objectives and programs of the late-Weimar regimes—ended the Depression; with the economic upturn, the profit picture improved significantly for the chemical trust. From 1933 to the beginning of 1936, the firm posted net profits of between 6 and 7 percent of turnover. Net earnings exceeded 10 percent in 1936, the first year of the Four Year Plan, and came to nearly 14 percent in 1940. During the same period, turnover almost doubled. Even in the closing years of the war, profits stayed high, amounting to more than 9 percent in 1942 and 1943, and 5.7 percent in 1944.

Profits posted during the Nazi period are deceptively low. Research budgets increased in absolute terms, though not in percent of turnover. Retained earnings and write-offs permitted extensive investment in new plants, although not always in geographic or production areas favored by the firm's leadership. Once the war was underway, Farben profited in still a third way: it took over the chemical factories of competitors in countries overrun by German armies and/or forced them to sign contracts advantageous to the I.G.[64]

Three factors explain the firm's enormous success in the Nazi period: its ability to deliver goods necessary to achieving Hitler's objectives; its prominent role in the Nazi regime; and, its willingness to "howl with the wolves." I.G. and the Nazis shared aims in the production of *ersatz* materials, in particular synthetic oil, rubber, and nitric acid. In the long term, the war the Germans embarked upon in September 1939 would not have been possible were it not for production of synthetics. As one of its component firms had in World War I, when synthetic nitrogen enabled Germany to survive four grueling years of war, the I.G.'s synthetics programs allowed the country in the short run to transcend its resource limitations. Profits from the synfuels program were disappointing,[65] but they helped offset the ill-advised investment in its development incurred during the 1920s. Profits from other areas more than made up for poor performance from the synthetic fuels sector.

Farben personnel played a prominent and vital role in the Nazi regime. Carl Krauch, for instance, headed the high-pressure chemistry division of the company (Sparte I) until 1938. From 1940 until the end of the war, he was chairman of the supervisory board. At the same time, Krauch occupied important positions within the Nazi state apparatus. He helped formulate the regime's Four Year Plans and the Karinhall Plan. The plans all benefited the chemical industry in general and I.G. Farben in particular. He worked to implement the programs as head of research and development for the chemical industry within the Raw Materials and Foreign Exchange Staff (the forerunner of the Four Year Plan organization), and retained a similar position in the later organization. In 1938, he was appointed commissioner-general for problems of the chemical industry.

Another member of the I.G. managing board, Heinrich Bütefisch, was involved both in the company's direction and in the state apparatus. Bütefisch was the director of the Ammoniakwerke-Merseburg, or Leuna Works, which was the largest I.G. synthesis facility. As chief of the Economic Group of the Energy Industry within the Economics Ministry, Bütefisch helped to control, coordinate, and regulate the fuel industry. He acted, in addition, as head of the Working Committee for Hydrogenation, Synthesis, and Carbonization, which advised the government regulatory hierarchy on the synthetic oil industry, and coordinated government direction and industry production. A further connection existed between Bütefisch and the Nazi party and regime: since he belonged to the Nazi *Freundeskreis der Wirtschaft* (Circle of Friends of the Economy), he often attended party functions as the special guest of Wilhelm Keppler and Heinrich Himmler. Krauch and Bütefisch were only the most prominent of a number of I.G. executives who also played some role in government during the Nazi period.[66]

Complicity in the practices of the Nazi regime did not stop at coincidence of production objectives or participation in administering Germany: the I.G. employed slave labor in some of its plants. More specifically, the firm erected a facility near the site of the Auschwitz concentration camp in order to make use of slave labor. Its leaders continued construction and, eventually, operation of the plant—even when evidence mounted of what was occurring in the main concentration camp and defied attempts to overlook it: at the very least, the stench of the incinerators must have brought questions to the minds of visiting I.G. executives.[67] Yet they refused to draw any conclusions from such evidence, let alone to attempt (even though such an attempt would probably be in vain) to do anything to stop the practices. Of all the alleged and actual crimes of the giant chemical corporation, this proven instance of complicity in the worst deeds of the Nazi regime demonstrated that the practices and objectives of the party and those of the firm, while not identical, nevertheless coincided to a disastrous degree.

There are mitigating factors in the story of I.G.'s involvement with the Nazis. For instance, the dual duties of Carl Krauch in the firm and in the regime led to a conflict of loyalties, with the allegiance to the regime and its aims quickly gaining the upper hand.[68] Farben thus did not gain as much as one would expect from Krauch's participation in the governmental apparatus. On the other hand, there were numerous other I.G. Farben executives seconded to government service, and it is not clear that Krauch was representative of the group. Heinrich Bütefisch, for instance, may have been more typical. Bütefisch was involved with dealings with the party rather early, meeting with Hitler even before the seizure of power. His position as head of the Leuna works meant that he gained substantially from fulfilling Nazi economic plans to the hilt. He had, furthermore, close contacts with the head of the SS, Heinrich Himmler.[69]

In addition, it is important to emphasize that the I.G. may not be considered as monolithic in any evaluation of its relations with the regime. In fact, just as certain industries gained more (if only in the short term) from Nazi economic policy than others, and certain firms within industries gained more than other firms,[70] certain divisions within a particular firm gained more from Nazi war preparations than others. Within the I.G., marked differences in attitude toward and relations with the regime prevailed, depending upon the product area(s) of a given division. Personnel from the high-pressure chemistry division, and from plants heavily involved in such production, often had good, if not magnificent, relations with the regime; personnel from divisions of less interest to the regime's planners fared poorly.[71]

In summary, it is clear that the regime determined the political aims during the Nazi period and then asked firms to provide the means to fulfill them. Nazi policies of autarky, and their preparation for and implementation of an aggressive war of expansion to gain *Lebensraum* in the east, set the tone for the era.[72] But the I.G., because of its technical expertise, production concentration and capacity, and financial might, shared in the interests and objectives of the regime to an extent unparalleled among large firms. These strengths, combined with the I.G.'s export earnings, afforded the firm a certain independence and freedom of movement uncharacteristic of German firms. For this reason, more nuanced evaluations of the role of the I.G. under the Nazis should not obscure a number of crucial facts: that representatives of the I.G. participated actively in the regime from at least 1936; that the company's leaders used their position to take advantage of German battlefield victories to subjugate much of the European chemical industry to its control; that they utilized slave labor at production installations such as Auschwitz; and that they turned a blind eye to the mass murder carried out by the Nazi regime in the concentration camp near I.G.'s Auschwitz plant. Certainly by the late 1930s and early 1940s, the firm came to accept and even to embrace some of the worst aspects of a monstrous regime.

Starting in mid-1940s, at the height of Nazi hegemony in Europe, I.G. Farben and other firms developed plans outlining their role in the Nazi New Order.[73] At the request of Gustav Schlotterer from the trade section of the Economics Ministry, Farben prepared a program that foresaw the creation of a European bloc for chemical trade. I.G. Farbenindustrie A.G. would dominate the trading area by taking over significant shares in ownership of the continental chemical industry, by obtaining favorable contracts for sales of organic chemicals in European markets, and by monopolizing for all practical purposes European exports of chemicals.

I.G.'s New Order planning marked the company's formal departure from the traditional orientation of the German chemical industry. From its inception, the industry had permitted some to dream of autarky since chemical firms produced goods that enabled transcendence of the country's poor natural resource endowment. But chemical industrialists, fully aware of their dependence on overseas markets, in general favored free trade and limited state intervention in the economy. During the Nazi period, Farben's management gradually abandoned its earlier beliefs, partly out of necessity (since the Depression disrupted world trade, hurting the I.G. more than many other firms) and partly out of desire for high profits. The New Order plans illustrated clearly the final conversion of the firm's leadership from economic liberalism to neomercantilism.[74] The plans, along with the trust's use of slave labor and its unwillingness to recognize, criticize, or stop the regime's mass murder program, exemplified Farben's gradual and fatal emulation of Nazi policy and practice.

According to numerous commentators, at about the same time as the New Order plans appeared (mid to late 1940), I.G. leaders began to discuss another program for the postwar period: *Selbstentflechtung*, or the self-imposed division of the company into a number of smaller ones for the purpose of better efficiency.[75] If one could link these discussions with postwar developments (i.e., the Allied breakup of the firm), one might show that the earlier plans to break up the I.G. were substantially adopted in the Allied disentanglement of the corporation, and/or that the persons responsible for developing such plans during the war acted as advisers to the Allies or German government officials in the actual decision making surrounding the implementation of *Entflechtung* (breakup of the firm). Based on available evidence, there was no such linkage. Comments on plans for *Selbstentflechtung* within the I.G. in the early 1940s were just so much pious talk. Members of the managing board undoubtedly lamented at times the cumbersome structure of the firm. But they did nothing to plan for, let alone implement, an internal breakup of the I.G. during the war.

Before long, however, in the wake of Allied successes, New Order plans and the empty talk of *Selbstentflechtung* gave way to programs based not on assumptions of German victory, but rather of German defeat. The word "program" is probably overly formal: there was some scrambling within the firm beginning in 1942–1943 to try to minimize the effects of Allied policy in the event of an increasingly likely Allied victory. Top I.G. managers tried, for instance, to ensure continuity in the event of a German defeat. To this end, the leadership appointed the young Ulrich Haberland director of the Bayer group of factories in 1943. He was, however, *not* appointed to the managing board at the same time, despite the fact that the Bayer director normally attained such a position *ex officio*, for the express reason that the Allies were not likely to remove him from his position if he were not a board member. Instead he would serve as a focal point in the reconstruction of the company.[76] I.G. management could not, of course, predict that occupation authorities would retain Haberland in his position. Still, the tactic demonstrated that the firm attempted to second-guess the Allies, and, in this case, the move paid off.

I.G. representatives also attempted to avert the worst consequences of an Allied victory through transfer abroad of patents and other assets. A minister appointed by the provisional French government to represent its interests in Spain, for instance, noted in March 1945 the presence of a number of German "tourists" in Madrid and Barcelona. He continued:

> Among the most important visitors were the general director of the "I.G.-Farbenindustrie," a delegate of the exchange office in Berlin, some financial representatives, and some engineers.

The visitors had all spoken with Spanish bankers and with representatives of "the enterprises in Spain controlled by I.G. Farbenindustrie with their branches in Portugal and South America." The purpose of their discussions was clear to the minister: transfer of ownership of German assets to Spanish nationals in order to try to avoid confiscation by the Allies.[77] Similar actions took place in Switzerland and Sweden in the closing months of the war.

A further move to minimize the negative consequences of the projected Allied victory involved a more sophisticated attempt to anticipate Allied treatment of the chemical firm. In 1944, Hermann Schmitz, the chairman of the I.G. managing board, tried to apply to the I.G. itself the example of the recently completed breakup of the Wasag chemical firm into two component parts. One successor to the I.G. would hold all of the company's factories directly involved in war production. Schmitz expected to lose these facilities through Allied disarmament measures. The second firm would comprise all I.G. holdings unrelated (or only indirectly related) to war production. Schmitz thought that the Allies would retain these factories in Germany for peacetime production, and that the I.G. would eventually reassert its control over them. In effect, Schmitz developed a plan for *Selbstentflechtung*, although his scheme had more to do with desperation than with the dictates of efficiency or economy. He made preliminary arrangements for financing the breakup, but for some reason the plan was dropped before it ever came to fruition.[78]

In summary, then, there is no solid evidence of extensive, detailed, or long-range planning for the postwar period within the I.G. Farben in the event of an Allied victory. Nor should we really expect it. The confusion of the conflict itself was not conducive to detailed consideration of options for the postwar period. In addition, postwar planning on the basis of a projected Allied victory would, at best, have smacked of defeatism. At worst, it would have been considered high treason. Therefore, no strategic planning for the reconstruction took place. Those actions embraced involved tactical maneuvering to make the best of a rapidly deteriorating situation.

In May 1944, concentrated bombing attacks on German industry began. By June, U.S. General Carl A. Spaatz directed that the "primary strategic aim of the U.S. Strategic Air Forces is now to deny oil to enemy air forces."[79] Since the German chemical industry, and in particular the I.G. Farbenindustrie A.G., produced much of the Third Reich's oil, it soon felt the effects of the new strategy. Bombs directed at synthetic oil production facilities often landed on other production facilities within a plant, or, just as importantly, disturbed the flow of raw and intermediate materials within, or to and from, the factory. In the highly interdependent chemical industry such a disruption could mean only one thing: rapid production loss.[80] By the beginning of 1945, production in the German chemical industry was limping along, and the I.G. was but a shadow of its former self. From the point of view of the leadership of the firm, the worst was yet to come.

PART II
THE I.G. IN THE INITIAL OCCUPATION PERIOD

With the defeat of Nazi Germany in April/May 1945, Allied troops occupied the country. Each of the four major Allies—the Americans, the Russians, the British, and the French—took control over a strictly delineated portion of the country. Although the Allies had agreed to treat Germany as an economic unit, they divided it into four airtight areas, each with its own set of cultural, social, political, and economic policies.

For I.G. Farben, the advent of the occupation marked the end of an era. Its headquarters building in Frankfurt was seized, first by displaced persons and then by the U.S. Army. By the fall of 1945, all of the Allies had agreed in principle to break up the firm. What is more, at least in the short term, production relationships among the corporation's factories were disrupted. This was extremely significant for the I.G., which had by and large constructed its works groups on product rather than regional lines. Finally, research, the cornerstone of the I.G.'s success, suffered during the early occupation. Already hard hit by cutbacks during the Nazi period, researchers after 1945 were encumbered with Allied restrictions. Even when the restrictions were gradually lifted in the late 1940s and early 1950s, efforts tended to be devoted to applications of existing knowledge rather than basic research.

Three of the headquarters of the I.G. works groups lay in the western zones of occupation, each located in a different zone. Thus, attention to the three main I.G. successors in the immediate postwar period allows an opportunity to investigate not only the impact of the occupation on production relationships between the factories, but also the effects of differing policies on the successors. The following three chapters concentrate on these issues in each of the western zones of occupation, giving at the same time an in-depth portrait of the history of the Hoechst, Bayer, and BASF works groups.

2
RADICAL DECONCENTRATION AND ITS RESULTS: THE AMERICAN OCCUPATION AND THE HOECHST GROUP 1945–1946

In a popular characterization of the four zones of occupation in postwar Germany, the Russians had the farmland; the British had the industry; the French had the wine; and the Americans were left with the scenery. The U.S. zone of occupation, comprising a newly created "Greater Hesse," parts of the eventual province of Baden-Württemberg, Bavaria, and the city-state of Bremen, featured some spectacular scenery. Yet the zone benefited from more substantial advantages, too. For one thing, heavily industrialized areas around Munich, Stuttgart, Mannheim, and Frankfurt, though heavily damaged by bombing, retained tremendous industrial potential: many factories remained surprisingly intact. Then again, some firms, most notably Siemens, moved their operations to the zone even before the end of the war. Later, as the conflict drew to an end, several important scientists and technicians avoided the oncoming Red Army by transferring their laboratories to what would become the U.S. zone. Finally, Frankfurt had been and continued to serve as an important financial center in Germany.[1]

In 1936, the (future) U.S. zone's chemical industry constituted a significant segment of total German chemical production: a postwar study estimated the net production worth of the chemical and chemical–technical sectors (excluding rubber and asbestos manufacture) at RM 2276 million. Of the total, the area that subsequently comprised the U.S. zone manufactured goods valued at RM 459.7 million, or 20.2 percent of the total. Chemical production in the U.S. zone of occupation was also more widely dispersed than in the other western zones of occupation. The American-controlled portion of the later-formed *Land* of Baden-Württemberg contributed 24.5 percent of the zonal production of chemicals; Bavaria produced 24.4 percent; Hesse manufactured 51 percent.[2] Thus, although it functioned as the hub of the industry in the U.S. zone, Hesse did not dominate zonal chemical production by the same margin as North Rhine Westphalia in the British zone or the Ludwigshafen area in the French zone. Chemicals constituted by far the largest single industrial group for the Hessian industrial economy as a whole, with 21.2 percent of total production. In the combined area (the British and U.S. zones after January 1, 1947), chemical manufacturing amounted to just 14.4 percent of total industrial production.[3]

One of the great works groups, or *Betriebsgemeinschaften*, of the I.G. Farbenindustrie A.G. centered around the "Hoechst Group" (officially known as the Middle Rhine, or Maingau Group), which was located in Frankfurt-Hoechst in Hesse. Frankfurt am Main itself was the corporate seat of the I.G., and, soon after the war, the firm's central office building on the city's Grüneburgplatz became the headquarters of the U.S. Army in Europe. But there was broader significance to the fact that the home office of the corporation fell in the U.S. zone: banks in the area held many of the I.G.'s liquid assets; and, the corporation stored many of its central records in and around Frankfurt.[4]

When U.S. troops occupied their zone in the late spring of 1945, American policy[5] dictated that they concern themselves extensively with the zone's chemical production facilities. During the early months of the occupation, U.S. officials seized control of chemical factories, kept watch over their production and research, and took action against their managers. Later, as American authorities recognized apparent economic and political limitations to their actions, Germans from governmental agencies and from industry played a key role in policy formation and change.

U.S. POLICY DEVELOPMENT, 1944–1945

Even before the active involvement of the United States in World War II, the prospective shape of the postwar world occupied government policymakers.[6] U.S. combat troops became involved in the conflict on an active and massive scale in 1942 and 1943, and interest of high-level American officials in the subject grew. When the war appeared to be drawing to a close on the European front in 1944, they began concentrated deliberations on the postwar order.

Specific recommendations on policy toward Germany in the postwar era emanated first and foremost, surprisingly enough, from the U.S. Treasury Department. Treasury secretary Henry Morgenthau, Jr. parlayed his special position as friend and neighbor of President Franklin D. Roosevelt into bureaucratic ascendancy within the Roosevelt administration. Morgenthau and a handpicked group of expert advisers took the lead in planning for the postwar international financial order and for the treatment of Germany. The so-called Morgenthau Plan was approved by Roosevelt and British prime minister Winston Churchill in Quebec in September 1944. In calling for deindustrialization and disarmament of the enemy nation, the Plan proposed quashing the might of the Ruhr district—Germany's industrial heart—by flooding its mines and dismantling or destroying much of its industry. The former industrial giant was to be transformed into a predominantly agricultural country. Adverse publicity in the United States immediately after the endorsement of the plan in Quebec caused Roosevelt, who was campaigning for reelection, to withdraw his support for it.[7]

Beginning in April 1945, American planning for the German occupation continued under the auspices of the Informal Policy Committee on Germany (IPCOG). Representatives of major governmental departments and agencies—including the Departments of Treasury, State, Navy, and War, and the Foreign Economic Administration—constituted IPCOG's membership. Morgenthau's personality dominated its meetings, but they also reflected the president's rejection of his plan and the special position of the War Department in the impending occupation. The Army would, after all, be the organization responsible for actual implementation of U.S. government instructions.[8] President Roosevelt's death on April 12, 1945, and his replacement by Vice-President Harry S. Truman further undermined Morgenthau's power. Thus, IPCOG's final directive, approved by the new President and dispatched to General Dwight D. Eisenhower on May 16, 1945, contained Morgenthauian provisions. At the same time, it showed consideration for the concerns of the War Department: to keep the occupation as short and as inexpensive as possible; and, to ensure fulfillment of army responsibilities to supply both the occupying forces and Europe's civilian population.

JCS 1067, as the directive from the Joint Chiefs of Staff was known, ordered the U.S. Military Governor in Germany to move to disarm, deindustrialize, and denazify the defeated nation. Nevertheless, he retained autonomy in his zone, could proceed in the short term with the rehabilitation and operation of war-related industries such as synthetic oil and rubber, and was permitted to take any steps necessary to prevent "disease and unrest" among the German population.[9] Economic provisions were central to the U.S. policy directive. JCS 1067 instructed the military governor to do several things: he could take no steps whatsoever to rebuild the German economy (except those permitted under the "disease and unrest" formula); he was supposed to dismantle or destroy all industry directly related to war; he could allow the Germans to keep and use factories indirectly related to war, but only if the facilities did not represent "excess capacity." Groups of Allied experts had the responsibility of determining "excess capacity" and recommending a suitable level of German industry based on an evaluation of German war potential and an estimation of the minimum acceptable standard of living. Other Allied specialists would carefully supervise all industry left in Germany.[10]

Addressing only briefly policy on individual industrial sectors, the directive left such detailed measures to Level of Industry planners. The chemical industry's special status within the German war economy, however, necessitated singling it out several times. JCS 1067 specifically identified explosives manufacture and most synthetic processes (e.g., for the production of synthetic ammonia, rubber, and petroleum) as industries directly related to war potential. The military governor would have to liquidate them in the long term, although he could in the short term recommence production of synthetics for the purposes of the occupation. The directive also limited production in certain areas of the industry, and, by controlling and limiting German research, struck at one of the important bases of success of the prewar and wartime German chemical industry.[11]

JCS 1067 offered no guidance on key aspects of U.S. policy toward the German chemical industry. For instance, a separate set of instructions from various governmental agencies outlined the multifaceted American policy on technical exploitation. Even as the war was going on, the U.S. Alien Property Custodian moved in this direction by confiscating existing German patents for chemical (and other) products and processes and allowing U.S. firms to use them without royalties during the hostilities. After the war, U.S. diplomats participated in an effort to formulate an accord on German patents. The accord held that:

> all former wholly German-owned patents [seized by August 1, 1946] ...shall be dedicated to the public or placed in the public domain or continuously offered for licensing without royalty to the nationals of all Governments, parties to this Accord.

The signatory powers agreed to create a central office to administer and coordinate the implementation of the accord and to take measures to protect the property of bona fide German refugees and non-Germans. The countries signed the agreement in London on July 27, 1946 and it came into force on November 30.[12]

The effect of the permanent expropriation of German holders of patents held to mid-1946 in the United States, Great Britain, France, and numerous other countries is not clear. In February 1947, a German patent attorney claimed that "American experts estimate the worth of the confiscated German patents abroad to be several million dollars."[13] But patents alone do not a product make: vital information on the actual construction and functioning of production systems remained in the hands of dispossessed German patent holders, and could be sold later to those wanting to utilize the former patents. What is more, rapid changes within the industry meant that in a relatively short time newly developed and patentable products made up an increasing portion of the revenue of a given firm. In 1975, for instance, Bayer A.G. held that 40 percent of its turnover resulted from the sales of products developed in the previous fifteen years.[14] The 1950s, a decade in which the changeover from coal- to petroleum-based chemistry occurred in West Germany, featured even more rapid change.[15]

Allied exploitation efforts were not confined solely to the seizure of German patents. Teams of experts from U.S. industry, as well as their counterparts from other Allied or neutral countries, swarmed over German industrial facilities in the immediate aftermath of the war.[16] Chemical technology constituted an important focus of activity for the investigation groups,[17] with synthetic oil and rubber foremost among the subjects investigated.[18] Allied experts looked into anything new or noteworthy in German chemical production techniques or products. For example, they interrogated Dr. Karl Winnacker, director of research and of several production areas at the Hoechst plant, on everything under his control. They also confiscated technical documents from Hoechst and published detailed descriptions of its processes in the areas of acetylene chemistry and plastics.[19] Within the chemical industry, I.G. Farben factories constituted a focus of Allied studies; U.S. authorities led the way. They had, after all, voluminous records at their disposal in Frankfurt and enjoyed a head-start on their Allies in inquiries into Farben's organization and technology.[20]

Technical exploitation teams presented their results in two different forms: general reports on products, techniques, plants, or plant personnel; and, microfilms or photocopies of technical documents. The latter often grew into very large collections indeed. Microfilms of technical documents on synthetic fuels, for example, amounted to well over 300 reels.[21] Although the materials were made available to U.S. industry, the extent to which it used them is questionable.[22] In their technical exploitation programs in the initial occupation period, American governmental agencies tried, through patent seizures and through technical investigations, to squeeze German industry for the sake of American industry. Their efforts probably had little impact on the postwar development of either American or German industry.

Policy toward the I.G. Farbenindustrie A.G.—not mentioned explicitly in JCS 1067—constituted another major element of U.S. policy toward the German chemical industry. Yet, despite silence on it in the occupation directive, the I.G. represented an obsession of U.S. occupation policy planners: key provisions in the military governor's instructions, including the commitment to control of research, prohibition or restraint of production of key chemicals, and the express goal of eliminating excess concentration of economic power in Germany, affected the I.G. more immediately and directly than any other firm in the German chemical industry. The intensity of the U.S. attitude was in marked contrast to that of its Allies, who were in general much less concerned with the I.G. per se. Long-standing American preoccupations with, and legal action against, any firm that dominated an industry to the point of near monopoly (or any firms involved in agreements tending to fix prices, divide up markets, or otherwise restrict competition) were the basis for that intensity.[23]

Attention to the position of the I.G. in the world chemical industry, to its market position in the U.S., and to its relationships with American firms such as Du Pont and Standard Oil of New Jersey began in earnest under the Roosevelt administration in the mid-1930s. Under the leadership of Thurman Arnold, the Anti-Trust Division of the U.S. Department of Justice, and especially its Patent and Cartels Section, pursued investigations of the German chemical giant. Joseph Borkin, later employed in the prosecution of I.G. Farben executives at the Nuremberg trials, was one of those Justice Department investigators. During the 1930s, I.G. Farben became his "Moby Dick."[24] Later I.G. Farben Control Officers and/or Decartelization Chiefs for the United States, James S. Martin, Myron Maupin, Randolph Newman, and Grant Kelleher, all either hailed from the Justice Department or had close connections and sympathies with the Anti-Trust Division.[25]

The division did not restrict itself to unilateral action in the courts against the German chemical giant. For one thing, Anti-Trust personnel wrote or inspired books and articles directed against the firm. Borkin, for instance, coauthored a widely read book published during the war on *Germany's Master Plan* (1943).[26] Featuring an introduction by Thurman Arnold himself, the book argued that Germany had aimed throughout its history at world conquest. To accomplish its task, the country had to overcome shortages in vital materials through technological expertise. Cartels and monopolies developed and controlled the technology; foremost among them was the I.G. The authors suggested that the Allies should learn from history at the end of the war: they should draw the lesson that control of the I.G. and other monopolies and cartels was the same as—and the only means to—control of German ambitions throughout the world.[27]

Representatives of the Treasury Department were the natural allies of the Anti-Trust Division within the U.S. government and in military government in Germany. Since the Treasury men were committed to a liberal world economic order in which cartels and massive combinations of economic power would have no place, they too disapproved of Farben's practices. Morgenthau's subordinates sometimes attained high positions within army and occupation finance sections. Bernard Bernstein, for instance, was a former treasury official who became the chief financial officer of U.S. military government in Germany. In that position, he and a number of others seconded from the department wrote a major report on the activities of the I.G. during the war. It contained recommendations on the treatment of the firm in the postwar period.[28] Bernstein's report and other material produced by Treasury Department and/or Anti-Trust Division personnel during and after the war had some impact upon policy toward the I.G. since, initially, they constituted the only background information on the firm in the files of both U.S. and British chemical-control authorities.[29] In addition, some Treasury Department experts worked on the prosecution staff in the Nuremberg trials of I.G. executives.[30]

Fundamentally, the publicity- and policy-formation efforts of the personnel from the Anti-Trust Division and the Treasury Department had a straightforward message. I.G. Farben had been responsible for providing the Nazi war machine with the materials absolutely necessary for the conduct of war, including synthetic ammonia, rubber, nitric acid, and petroleum. These were all things with which Germany was not naturally endowed, and under normal conditions the country would have had to import them from overseas. I.G. Farben thus contributed to making Germany immune to the effects of an Allied blockade. Then too, the firm was thoroughly entangled in the Nazi regime and, later reports alleged, supported the Nazis even before they came to power. The firm's representatives, it was often maintained, were partly responsible for aiding the Nazis in their seizure of power in 1933.[31]

In the view of many of those who would be active in the U.S. military government in Germany, the combine was thus the epitome of evil. The root of the evil was I.G.'s dominance of the German chemical industry to the point of near monopoly: the trust controlled virtually all of the ammonia production, owned or operated a considerable portion of the country's synthetic fuel capacity, and was responsible for most German chemical exports. Absence of competition led the company to assume that its power, especially when harnessed to that of an expansionist regime, was without limits. Limitless power corrupted without limit: it led to such evils as the I.G. Farben plant at Auschwitz, and the provision, through an I.G. subsidiary, of Zyklon B, the gas used most often to exterminate concentration camp inmates.[32] A twofold solution would eliminate such an excessive concentration of economic power that led inevitably to evil: first, it would be necessary to try the firm's leaders before an Allied tribunal—something that occurred later at Nuremberg—and to purge all members of the Nazi party from the firm's middle and upper management; second, the Allies must pursue a policy of radical deconcentration of the German chemical industry beginning at the top—with I.G. Farben.[33]

In fact, the American analysis was based on several misconceptions. True, Farben had been indispensable to fulfilling Nazi war aims, and the firm had thoroughly disgraced itself in its associations with the regime. But the I.G.'s role in bringing the Nazis to power was limited at most.[34] In addition, American policy failed to consider carefully the firm's place in the German chemical industry and thus proposed blanket deconcentration of the company in all areas. As noted in chapter 1, the I.G. dominated the German chemical industry but not to the same degree in all sectors. A more differentiated and carefully thought out policy might have recommended the breakup of, say, the firm's virtual dyestuffs monopoly, but left its inorganic chemicals and rayon production facilities alone. American misconceptions in this regard are, however, understandable when one recalls the background of United States I.G. control authorities: for the most part, they had concerned themselves with the firm's foreign operations, which sold products and rights to utilize technology in areas in which the company was especially strong.[35] Consequently, they regarded the corporation as monopolizing the entire German chemical industry.

Nevertheless, the interpretation outlined above of the I.G.'s role in preparing and carrying out the war dominated the minds of Americans responsible for the firm's control; it continued to have strong support within U.S. military government into the 1950s.[36] Most striking was its extensive similarity to a typical Marxist reading of the role of I.G. Farben in the origins and expansion of the

Third Reich.[37] Both stressed the dangers inherent in monopoly capitalism and the I.G.'s history as one of the worst examples of what monopoly capitalism could lead to. Both emphasized the role of the I.G. not just within the Third Reich, but also in helping it come about. Crucial differences separated the views of those favoring radical deconcentration from the Marxist analysis, however, and understanding those differences is essential to comprehending the postwar development of U.S. policy toward the I.G. American antitrust sympathizers which did not put forth a critique of capitalism per se, but rather of "bad" capitalism. Legislation and impartial supervision to curb the abuses of monopolists and to ensure competition, *not* the fundamental reconstitution and restructuring of the capitalist system itself—these were the basic tenets of the antitrust creed. What is more, many American reformers favored business units big enough to be competitive but not big enough to have a monopoly position. Consequently, even among the trustbusters, there were those who favored *oligopoly* as opposed to either monopoly or the total splintering of a firm.[38] The debate within the U.S. military government, and within the U.S. government in Washington, was from the start chiefly a discussion over what constituted "competitive units" and how large they should be.

Factions in the U.S. military government offered differing solutions to the problems of restructuring German industry. Initially ascendant in American policy, the antitrust element gradually lost ground to another group in the occupation regime that represented different interests and a rival tradition. Numerous figures from the U.S. business and financial communities represented the alternative element; their official positions extended from the technical exploitation teams to U.S. control authorities at the *Land* or factory level. Some of them counted among the closest advisers to General Lucius Clay, the later military governor of the U.S. zone.[39] The most prominent member of the group was General William Draper, head of the Economics Division, who was closely associated with business interests in the U.S. and in Germany through his earlier work with the investment firm of Dillon, Read, and Co.

In his book *All Honorable Men* (1950), James Martin delivered a damning indictment of the role of Draper and men like him in the development of U.S. policy:

After two and a half years [of activity in U.S. military government], I came back from Germany quite well aware that I had been wrestling with a buzz saw. We had not been stopped in Germany by German business. We had been stopped by American business.

The forces that stopped us had operated from the United States but had not operated in the open. . . . [W]hatever it was that had topped us was not "the government." But it clearly had command of channels through which the government normally operates.[40]

Martin's analysis showed his immense sense of personal betrayal at the development of American policy as it tended gradually but inexorably away from radical trust-busting of the I.G. But he underestimated the conflicts and confusion within the circle of these representing "business interests." Especially at the beginning of the occupation, U.S. businessmen in fact pursued contradictory objectives. If there were those who favored resumption of business as usual and who admired the German manager and his accomplishments,[41] some businessmen wished simply to take advantage of Germany's position to improve the U.S. competitive position. Why else, for example, did U.S. business loan valuable personnel to participate in military-government control authorities, or, more telling, to form technical exploitation teams?

Frequently, representatives of the U.S. business community in Germany wanted also to punish German business for its transgressions. This was especially true for the I.G., which was implicated in some of the worst crimes of the Nazi regime. Even Draper, whom Martin took pains to criticize directly,[42] did not consistently represent the interests of the German chemical firm, even more than a year into the occupation. Commenting in late summer 1946 on a memo from Martin on a proposal to establish penicillin production facilities (based on American technology) at the I.G. Hoechst plant, for example, Draper came down hard against the interests of the I.G.:

> I will discuss the investment angle to put in the penicillin plant at Hoechst with the Legal Division, but in the meantime any attempt to create a monopoly position or to hold up any information to any other group for their commercial ends must be completely stamped out. Please take the necessary steps to do this. If necessary we will move the

personnel and equipment from Hoechst altogether and set it up as a separate company. If the Wacker [chemical company of Munich] people are still without information needed to carry on their experiments see that the Hoechst group turn such information over to them immediately. We are not going to assist in developing penicillin production here for the benefit of the German people and have some particular group, particularly any formerly connected with I.G. Farben, attempt to profit commercially.[43]

Conflict within U.S. military government thus existed not only between the "antitrust" and "big business" groups, but also within the big business group. Contradictory policy directives were often the result. In 1946 an incident in Hesse displayed such policy confusion in its full contours. Given the U.S. policy of putting German industry back on its feet to allow its zone to support itself, military government encouraged export. Occupation authorities cooperated with the Hessian *Land* regime in setting up an export exhibition in Wiesbaden which opened in October 1946.[44] Products of the I.G. Farben successors in Hesse, of course, figured prominently in the exhibition, but the instructions on what would be allowed at the exhibition from the I.G. successors were enlightening. The deputy director of the Hessian MG Economics Division wrote that, "The general outline and program for the participation of the former I.G. plants . . . have been noted and there is no objection on the part of this headquarters to the carrying out of this program as stated."

But he attached conditions to the participation of the former I.G. plants:

> As previously pointed out, it is desired that every possible effort should be made to avoid capitalizing upon or otherwise creating the impression of relationship to the former I.G. Farben industry. . . . In general, trade names and trade marks affiliated with the former I.G. Farben industry should be excluded.
>
> The products on display should be offered, where ever at all possible, as emanting from an independent plant . . .
>
> It is further considered undesirable to utilize brand new trade names, or trade marks. In view of the fact that the plants are now under Military Government control and the future status uncertain [sic].[45]

Former I.G. plants could neither use well-known I.G. trademarks to command high prices, nor could they try to develop new markets for high-markup, name-brand products of their own creation. At the same time, they were to contribute heavily to the U.S. zone export effort. American policy seemed clearly confused. The former Maingau Group of the I.G. was the primary focus of that confused control.

THE HOECHST GROUP OF THE I.G. FARBENINDUSTRIE A.G.

The most salient feature of the chemical industry in the American zone of occupation was its relatively even dispersion, which was in stark contrast to its counterparts in the other western zones. I.G. Farben factories were widely represented in the zone, but they were not so widely spread out: of the forty-two major I.G. manufacturing facilities (i.e. with a hundred or more employees) in the American area,[46] most major plants owned directly by the I.G. were located in Hesse. The precise distribution of the corporation among the occupation zones is difficult to determine, since the number of properties was large and varied, including everything from dye factories to building societies and health insurance firms. In addition, Farben managed and operated a number of properties owned by the Reich, including synthetic oil plants and explosives factories. In October 1945, though, a study by the U.S. member of the Allied Control Council Coordinating Committee (CORC) estimated that "approximately 50 percent of the I.G. properties are in the Russian Zone, possibly 10% in the French Zone, and the remaining 40% is more or less equally divided between the American and British Zones."[47] Thus, about 20 percent of the former I.G.'s plants and assets overall were in the U.S. zone.

One factory complex stood out above all the others in terms of its administrative, production, and historical importance. The I.G. facility located at Hoechst near Frankfurt was the central factory, or *Stammwerk*, of the I.G. Maingau Works Group: its leadership directed the production facilities of one of the former I.G.'s main production groups. Historically, the Hoechst group developed similarly to the other major western-zone I.G. groups that centered around Bayer and BASF.[48] Founded in 1863 close to the site of the Hoechst *Stammwerk* by two salesmen, Wilhelm Meister and August Müller, and two chemists, Dr. Eugen Lucius and Dr. Adolph Brüning, the Meister Lucius & Brüning company started out on a small scale: just five workers, a clerk, and a chemist manufactured and sold aniline dye products. Aggressive sales enabled rapid expansion, especially into foreign markets. In the mid-1860s the company captured fashion attention in France when the Empress Eugénie wore a green silk dress dyed with a newly developed Hoechst product. By providing dyes for the textile industry, the firm entered the English market.

In the 1880s, Meister Lucius & Brüning also moved into other product areas, especially pharmaceuticals. New markets and products allowed a sharp increase in personnel, to 370 workers in 1879, and to 1860 workers in 1888. A growing stress on research accompanied the expansion: Robert Koch developed his diphtheria serum in conjunction with the company, while Paul Ehrlich sold it the rights to produce Salvarsan. In-house research increased, and the number of chemists grew to twelve in 1879 and to fifty-seven in 1888. The number of engineers and technical administrators increased from none in 1879 to nine in 1888.

Adopting strategies similar to those of other large firms, the Hoechst firm participated in the rationalization movement of the German chemical industry in the early twentieth century. It enlarged its own facilities still further, took over smaller firms, and made agreements to limit production and to divide up markets. The company played a major role in the *Dreiverband*, formed with Casella and Kalle between 1904 and 1907, and joined the Little I.G. in 1916. By the end of World War I, the Hoechst-based company had sales agents throughout the world and offered a full range of chemical products. With the fusion that created the I.G. in 1925, the factory became the center of the Middle Rhine production group. The Hoechst works supervised and controlled production at the I.G. plants at Gershofen, Knapsack, Marburg (Behringwerke), Mainkur (Casella), Offenbach, and Griesheim (two separate facilities). The group was thus heavily concentrated in and around Frankfurt in what would become the U.S. zone of occupation.[49]

Like others in the I.G., the Middle Rhine group produced a broad palette of chemicals. The single main factory at Hoechst produced the widest variety of products, and it was closely connected in terms of production with the other, more or less specialized factories in the group.⁵⁰ One thing that differentiated the group technologically from those based at Leverkusen and Ludwigshafen, however, was its reliance on outside sources for key materials basic to the production process. For instance, the most important department at the main plant was for producing sulphuric acid, on which most other departments relied directly or indirectly. Pyrites, necessary for sulphuric acid production, came from Spain, Cyprus, and Norway until the late 1930s. During the war and in the immediate postwar years, pyrite supplies came from a mine in Meggen, about seventy miles east of Düsseldorf in what became the British zone of occupation. Production of nitric acid, also the basis for an array of other Hoechst manufactures, depended upon supplies of synthetic ammonia from Oppau (in the Upper Rhine Group). The third most important set of preliminary products at Hoechst, hydrochloric acid and chlorine, depended on hydrogen (obtained from Ruhr coke-oven gas), power (which also originated from the Ruhr), and tar (from Bitterfeld in the Russian zone of occupation).⁵¹

In business terms, the I.G. era—including the Nazi period—was not a particularly positive one for the Hoechst group. Lacking top managers of the caliber of Bosch and Duisberg, Hoechst and its associated plants could not keep up with their counterparts in Ludwigshafen, Leverkusen, and Middle Germany. Investment in new plant was relatively neglected,⁵² research facilities became antiquated, and substantial production lines—including some dyes and other basic products—were lost to the other I.G. groups.⁵³ The group continued to manufacture pharmaceuticals, especially at the Behringwerke, but significantly they all bore a single "I.G." trademark, the Bayer cross.⁵⁴ The location of the main plants along the Main River constituted an additional problem because of its relatively high summer temperatures.⁵⁵

As a former wartime manager Karl Winnacker—and later (1951) chairman of the Hoechst management board—noted, the result of all these drawbacks was that

> the technical preconditions were much more unfavorable here [at Hoechst] than at Bayer or BASF. Transportation and energy costs, in short everything that went into the calculation [of prices] lay much higher at Hoechst than at other works. The splintering-up of antiquated production facilities [Hoechst featured a number of geographically close but not contiguous plants] was responsible for this.⁵⁶

Kurt Lanz, who would later help reconstruct Hoechst's export trade, characterized the group as "a sleeping giant" when he joined it in 1937.[57]

If the group received poor treatment under the I.G., though, its plant equipment was also relatively intact after the end of hostilities. Gauging the exact extent of damages to the various plants poses difficulties. But information available on two of the factories associated with the group indicates just how limited destruction was. In September 1944, two bomb attacks struck the Kalle & Co. plant in Wiesbaden-Biebrich, an "independent" facility under the control of the Hoechst group. Production resumed following repairs, and, despite more bombing attacks in the first months of 1945, "the plant is today [June 11, 1945] totally or at least in most cases ready to work."[58] Damage to the main plant itself proved even more limited. The Production Control Agency (an arm of the Allied Supreme Command) investigated the plant at Hoechst and determined that building damages amounted to just 5 percent—which probably meant that the windows had been broken and some of the roofs had collapsed from the force of bomb blasts and in the final fighting of the war. Machinery and equipment for production in the plant had only "small damages." There was "increasing loss, however, by corrosion during prolonged stoppages."[59]

Because of its forward position on the western front, U.S. troops occupied the main plant of the Hoechst group by March 28, 1945. A report by Lt. Col. Herbert G. Moulton on July 30, 1945, gave a clear and comprehensive picture of the condition of the Hoechst *Stammwerk* in the early months of the occupation. Besides reemphasizing the conclusions of the Production Control Agency that there was virtually no bombing damage, Moulton noted that the factory produced few materials directly related to the war effort. In terms of its equipment, the plant was unremarkable, with much of the hardware old-fashioned or simply worn out. Hoechst depended almost entirely upon outside sources of supply for raw materials and energy. Thus, although the physical plant was only slightly damaged, Moulton pointed out that "the value of the plant is in its operating organization rather than its physical assets." He consequently recommended that any prospective Allied policy on the plant concentrate upon its organization and human resources rather than upon the physical facilities themselves.[60] U.S. policy would, in fact, have dealings with all three.

U.S. POLICY IN ACTION: THE HOECHST GROUP, 1945–1946

U.S. policy toward I.G. Farben plants in its zone differed from that of the other western Allies both in the speed with which it was implemented and in its severity. The root of the differences lay in the strength of feeling concerning decartelization, deconcentration, and denazification within the U.S. military government.

Armed with the results of wartime Anti-Trust Division investigations of the I.G., in possession of the firm's headquarters building in Frankfurt, and heavily staffed with "decartelizers," the U.S. military government was first among the Allies to take formal action against the firm. By General Order No. 2 pursuant to Military Government Law No. 52, the American authorities seized direction and control of the properties and assets of I.G. Farbenindustrie A.G. in their zone. The order directed the deputy military governor to carry out U.S. policy objectives with regard to the chemical industry, including making available plant equipment for reparations, destroying war-related production facilities, and "dispersion of the ownership and control of such of the plants and equipment seized under this order as have not been transferred or destroyed." It suspended the rights of I.G. shareholders and removed the managing and supervisory boards from control of the company. The basis of such extreme measures, made explicit in the directive, reflected the argumentation of decartelizers: I.G. Farben, by virtue of its cartel relationships and its central role in "building up and maintaining the German war machine," was to be split up. The order took effect on July 5, 1945.[61] In the following months, the other Allies took formal action on the properties of the I.G.; a four-power plan came only in late November 1945.

Severity was another hallmark of U.S. action on the I.G. Virtually from the beginning of the occupation, Farben's plants in the U.S. zone were forced to be self-reliant, independent entities. American officials from the very start severed technical and commercial relations among the factories of the Hoechst group. The experience set the Hoechst group apart from that of BASF, whose two main plants at Ludwigshafen and Oppau operated in tandem under French control, and that of the four Lower Rhine group factories in the British zone, which continued for all practical purposes to operate as a single production unit.

I.G. Farben control authorities in the U.S. zone also applied denazification statutes more strictly than the other Allies. Unlike the British zone, where Bayer and its satellite factories operated without interruption under the leadership of Ulrich Haberland, American officials removed managers of the Hoechst group factories who had been associated with the Nazi party. In the case of Karl Winnacker, wartime director of most chemical production and research at the main Hoechst plant and later chairman of Hoechst A.G., the U.S. action involved first an undesired stint as a fruit picker and then the directorship of the Knapsack plant in the British zone. He attained the position at Knapsack through the intercession of Haberland. Winnacker's experience of manual labor and/or some unemployment was repeated frequently among the Hoechst group's top managers (and among managers of other U.S. zone firms); that meant that the various plants of the group were often left with inexperienced personnel in top positions.[62]

In August 1947, pursuant to an indictment handed down the preceding May, U.S. prosecutors began a trial of important I.G. managers at Nuremberg. Although the action occurred much later than the events described so far, it is indicative of the nature of the U.S. policy toward the chemical firm. Beginning well after the conclusion of the Allied (i.e., four-power) trials of major war criminals at Nuremberg, the I.G. trial was entirely under U.S. jurisdiction and responsibility. Prosecution attorneys prepared indictments for members of the firm's managing board, the chairman of the supervisory board (Carl Krauch), and a few persons selected from other key I.G. positions—a total of twenty-four men. Twenty-three were tried for various crimes (one of those charged became ill and was thus exempted from trial), including conspiracy to plan and carry out an aggressive war of expansion, spoliation, and crimes against humanity. This final charge related in part to the firm's use of slave labor at its plant in Auschwitz. Announced about one year later at the end of July 1948, the results of the trial bore astonishingly little relation to the alleged crimes: ten of the twenty-three men were fully acquitted, with the remainder sentenced to prison terms lasting from one and one-half to eight years. Even those convicted of use of slave labor and mass murder—five men in all—received relatively light sentences of six to eight years each. One could sympathize with chief prosecutor Josiah DuBois in his bitter assessment of the sentences as "light enough to please a chicken thief."[63]

Rapid and severe though it was, implementation of U.S. policy toward the I.G. Hoechst group lay in the hands of a bewildering array of organizations. At the highest level, the Supreme Headquarters of the Allied Expeditionary Force (SHAEF) was dissolved in July 1945. Many SHAEF personnel were transferred to the U.S. Group Control Council (US Group CC) in Berlin, or to the G-5 Division of the U.S. Forces, European Theatre (G-5, USFET) in Frankfurt and Hoechst. Some found employment in the organization formed in the fall of 1945 to take over the duties of both the US Group CC and G-5 USFET, that is the Office of Military Government for Germany, U.S. (OMGUS).[64] Others were sent to do duty in the Pacific theatre where the war continued until August 1945, while others still were demobilized quickly, especially when public sentiment in this direction grew in the late summer of 1945.

Adding to the state of flux in the U.S. organizational apparatus was the confusion created by the interaction of several factors: transportation and communications difficulties within Germany; shortages of vital raw materials and foodstuffs; hundreds of thousands of homeless civilians and displaced persons; and roving teams of Allied investigators. The chaos of the first few months of the occupation engendered organizational paralysis and overlapping authority throughout Germany. Existence of extreme confusion and activity at cross-purposes cannot be stressed enough in any evaluation of the initial occupation period.

Lines of responsibility within U.S. military government as concerned the I.G. were not clearly defined either; a number of divisions, and branches within divisions, were responsible for control and direction of the firm. The Finance Division controlled the corporation's properties and assets. The Chemical Branch of the Economics Division of OMGUS oversaw production quotas and technical matters for the chemical industry as a whole, including those of I.G. plants in the U.S. zone. For the Hoechst *Stammwerk* and for most of the plants of the Hoechst group, direct responsibility for production and technical matters devolved upon the Chemical Branch of the Office of Military Government (U.S.) for Hesse (OMGH). Much of the direct personal contact between U.S. occupation authorities, personnel from industry, and provincial and local authorities took place at this level. The Research Control Section of OMGUS and OMGH regulated the type and amount of scientific and technical research done in I.G. facilities.[65]

Supervision of dispersal and breakup of the I.G. in the U.S. zone, coordination of the many responsible military government agencies, and on-site administration of individual factories fell under the province of the Decartelization Branch of the Economics Division of OMGUS. The closely associated I.G. Farben Control Office, which was created on July 5, 1945, in conjunction with General Order No. 2 to seize the I.G., shared the Branch's responsibilities. The staffs of the two organizations included numerous "trustbusters." James Martin, himself a former Anti-Trust Division man, acted as head of the branch for most of the early occupation period. Martin took over the position of I.G. Farben Control Officer as well, and, according to an Economics Division memo, the dual responsibility was to continue. But his successors as Branch chief for the most part appointed separate control officers.[66]

The I.G. Control Office seized the property of the trust in July 1945 with a grand total of thirty-three officers, most of whom were seconded from the Finance, Legal, and Economics Divisions of U.S. military government. Since Control Office personnel were "on loan" and because of the size and complexity of the task of controlling the gigantic chemical firm, the initial arrangement was intended to be temporary. Thus, the Office requested a vast increase in its personnel roster; in September 1945, it obtained permission to expand to a total of 142 American personnel (including 61 officers).[67] It never attained the assigned number. Officers and civilians on loan from other branches returned to their former jobs or left Europe without replacements coming through. At the beginning of 1946, drastic staff reductions in OMGUS across the board meant still further cutbacks. Discrepancies between allocated and actual personnel in the Control Office continued. In June 1946, an allocated 42 personnel stood against an actual staff of 25. The number dwindled steadily and sharply to just 4 in mid-1949, and the Control Office got by with no stenographic personnel whatsoever from fall 1947 to August 1949. From the beginning of the occupation, the size of the staff was smaller than its counterparts in the British and French zones.[68]

Decreases in already low numbers of U.S. officials involved in supervising the chemical industry and the I.G. formed part of a broad-scale trend within U.S. military government in 1945 and 1946. They thus reflected an important paradoxical aspect of U.S. policy toward Germany in the immediate postwar period. On the one hand, U.S. moves became for a number of reasons more significant for western Allied policy than did French or British counterparts. Growing tension with the USSR was developing quickly into a cold war. At the same time, the French and especially the British faced the overwhelming domestic task of physical and economic reconstruction. Neither appeared equal to the challenge. The expense of occupying its zone of occupation in Germany put an even greater strain on the British budget and represented a further impediment to reconstruction.[69] The U.S. economy appeared to be the only one in the world strong enough simultaneously to handle its own needs and to aid in the solution of the massive economic problems faced by the Europeans. U.S. politicians and policymakers were at the forefront in identifying and acting against the perceived Soviet danger and were willing to use U.S. economic power in Germany and elsewhere to try to force the western Allies to go along with American foreign policy objectives.[70]

On the other hand, a key trend in U.S. policy as the occupation went on was to exert less and less direct control over the German economy. Instead, U.S. occupation officials moved quickly toward assuming a watchdog (i.e., supervisory) function with regard to the German economy. In this, they followed a tradition within the U.S. domestic government of loosely monitoring rather than directly controlling banking, industry, and other economic activities. The policy was thus one of strictly limited military government intervention. Numerous instances of the development of the U.S. stance might be mentioned. For instance, German supervisors and trustees who reported to central or *Land* occupation authorities gradually replaced American personnel in the supervision of individual German plants, including those of the I.G.[71] Direct responsibility for implementation of denazification policy also became a German task by June 1946. In general, turnover of direct policy implementation to Germans became more and more the rule beginning in early 1946.

GERMAN IMPACT ON U.S. POLICY, 1945-1946

As was the case in the British zone of occupation, Germans in the U.S. zone played from the start an important part in policy implementation. The two cases were quite different, however. Whereas a personality from the chemical industry such as Ulrich Haberland could influence developments in the British zone, chemical industrialists in the U.S. zone initially played a much more modest role. There were two reasons for this. First, the main Hoechst factory, the largest of the chemical plants in the U.S. zone, had no person of Haberland's abilities and reputation in the ranks of its top management. The shortage of talent was partly owing to a general shortage of extraordinarily qualified personnel in the group since before the I.G. fusion in 1925, and partly owing to strict U.S. denazification policies. Second, because component factories of the Hoechst group were treated as entirely separate firms after the beginning of the occupation, the leadership of the main factory wielded far less economic power than would otherwise have been the case. For these reasons, the German state apparatus in Hesse in the U.S. zone had a far greater hand in influencing occupation policy than Hessian chemical industrialists—and greater, too, than the state apparatus of North Rhine Westphalia in the British zone. As far as the I.G. was concerned, the Hessian Economics Ministry was quite important as well: in the middle of 1946, the other American zone *Länder* named it the main liaison to the I.G. Farben Control Office.[72]

"Greater Hesse," later simply Hesse, was a *Land* created after the war by the U.S. occupation authorities. It consisted of the former Reich provinces of Hessen-Nassau and Free Hesse and the *Bezirk* (county) of Kassel, minus some territory in the northwest of the province that was handed over to the French. Wiesbaden was chosen as the capital city of the new *Land*, since it was relatively intact and sufficiently large. As may well be imagined, destruction, a transient population, and a foreign occupation army, and the attendant confusion marked the first months after the end of hostilities. *Land* government had to be built from the ground up. Thus, even the *Landeswirtschaftsamt* (LWA), or *Land* Economic Office, which was erected on the basis of the former Reich LWA of Rhine-Main region, began to function fully only months after the end of

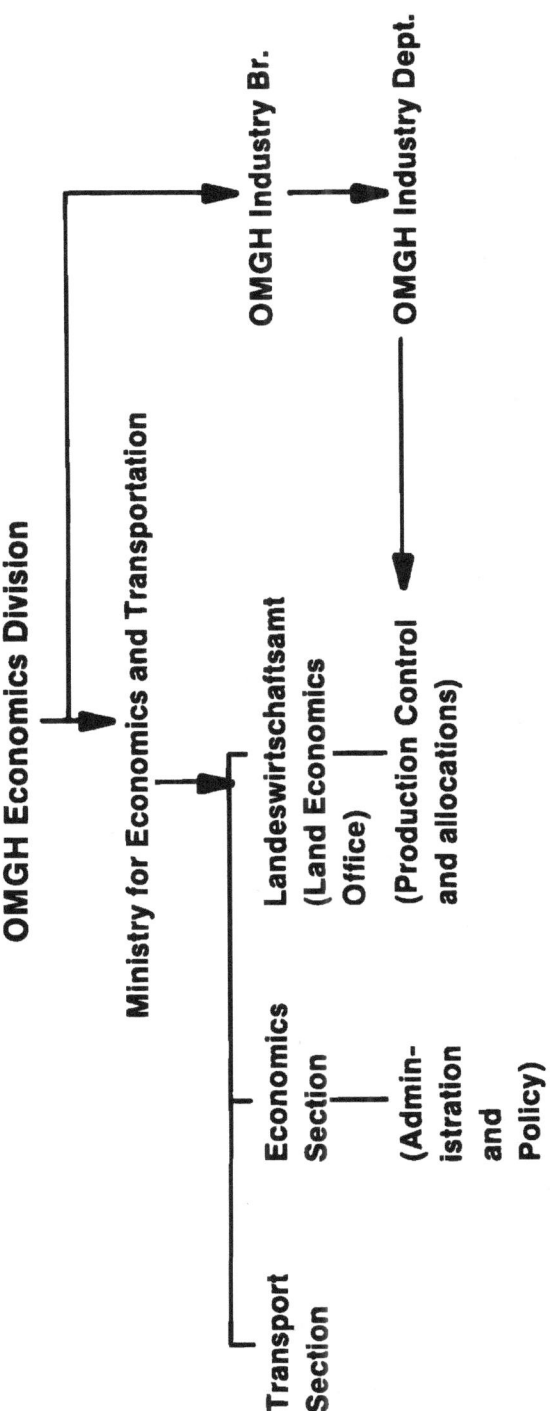

Figure 3: Government Economic Organizations in Greater Hesse (mid-1946).

[a] OMGH = Office of Military Government (U.S.) for Hesse.

Source: Compiled from Grosshess. Staatsmin. Der Minister für Wirtschaft und Verkehr. "Bericht über die Entwicklung der Wirtschaft im Lande Gr.-Hessen von Januar 1946–Ende Juni 1946." HStA Hessen, Wiesbaden, 507/4157; Economic Office of Land Greater Hessen., n.d. (ca. March 1946). HStA Hessen 649, OMGH ED Dir. Off. 8/82–2/4.

the war. LWA operations in Hesse at first restricted themselves to the city of Wiesbaden; after a few months, the LWA controlled allocation and production quotas in the *Bezirk* surrounding the city. Only toward the end of 1945 did it take over its duties for the entire *Land*.[73]

Figure 3 outlines the organization of the Hessian Ministry of Economics and Transport in a more stabilized form in July 1946, as well as the corresponding U.S. military government structure for the *Land*. The Hessian economic administration comprised two sections, the Main Economic Section, and the LWA. Responsibility for policy development for trade and industry belonged to the Main Economic Section, and it also had a number of supervisory functions including monitoring prices and research.[74] Day-to-day affairs connected with managing the Hessian economy were, however, the province of the LWA. It allocated important goods—including energy, machinery, consumer goods, bicycles, and so on—to individual firms. The LWA also coordinated production quotas and was responsible for the collection of statistics.[75] The LWA Industry Department thus served as the parallel organization to the Industry Branch of the OMGH Economics Division and employed a considerable staff for its tasks. In all, 72 of a total of 170 German civilians in the Economics Ministry were employees of the LWA Industry Department. Eighteen of 26 officers, 13 of 16 enlisted men, and 17 of 43 German civilians employed by the OMGH Economics Division were in the parallel Industry Branch as of March 1946.[76]

Several strong and influential personalities worked within the Hessian Economics Ministry. Dr. Lothar Wilhelmi, for instance, acted as head of the LWA Industry Department in 1946. He later became the direct assistant of the minister responsible for coordination of policy within the Economics Ministry, within the Hessian *Land* government, and with the bizonal and military government agencies.[77] In between these assignments, he and his colleagues in the Industry Department arranged numerous meetings with industry personnel. They brought together state and industrial figures in order better to coordinate policy and to present a united front vis-à-vis the Allies. In one such meeting, a dispute between the Chemische Werke Albert and Hoechst over who would produce superphosphate was decided (in favor of the former).[78] Early in 1947, his section helped resolve problems at the Griesheim works of the former I.G. in allocation of raw materials and represented the interests of the Hoechst *Stammwerk* against those of OMGUS in allocation of factory space.[79]

Dr. Friedrich Frowein, another key personality within the Hessian Economics Ministry in the initial occupation, had in fact been an employee of the I.G.; he worked for the firm as a research chemist between 1927 and 1932. From 1937 to 1945, he served in the *Dienststelle* Krauch (headed by Carl Krauch, head of the I.G. supervisory board from 1940 to 1945) in the German government, which was in effect the chemical industry section of the Nazi Four Year Plan Office. Interned initially as a suspected Nazi because of his affiliation with the *Dienststelle*, Frowein was released in April 1946; he was, in the words of an American investigator, "not a NSDAP member. Because of his cool attitude towards the Party he was excluded from promotions, which would have been commensurate with positions held. Subject is a known Anti Nazi."[80] Frowein became head of the Research Control Section of the Hessian Economics Ministry shortly after its creation in mid-1946 and remained there into the 1950s.

From the beginning of his tenure, Frowein had a good working relationship with his OMGH counterpart, Harry D. Coster. This was in part a matter of necessity for Coster since he was forced to turn over policy implementation to the Germans: he was the only person in OMGH involved in research control. But already in October 1946, Frowein wrote to Dr. Rudolph Mueller, the economics minister, praising the virtues of cooperation:

Practice in Greater Hessen has . . . shown that through close cooperation with the military government continuous opportunities exist to help research. Thus I have been successful in bringing to the attention of the American offices responsible for commercial legal protection [*gewerblichen Rechtschutz*] the connections between the duty to make commercial information public under [Allied Control Council] Law No. 25 [on the control of research] and the demands of secrecy and protection through patent legislation.

I have also been successful in bringing interest in German research to such a point that now Berlin has promised that seven copies of the most important specialized literature since 1940 will be made available in the U.S. zone through the American Chemical Society.[81]

Frowein's promotional efforts included attempts to gain permission for prominent professional journals in Hesse to resume publication. The editors of the world-renowned German *Handbook of Chemistry*, known as the "*Beilstein Redaktion*," were permitted to continue their activities under OMGUS auspices on the basis of an agreement between U.S. and German research control authorities.[82] Later, Coster and Frowein combined forces to ensure that the respected German journal, *Angewandte Chemie*, could retain sufficient office equipment to allow its publication.[83]

Cooperation between Coster and Frowein also aided in the expansion of research in individual I.G. Farben factories. The leadership of the Hoechst *Stammwerk*, for instance, received permission to increase its research projects from forty-three in 1946–1947 to sixty-nine in 1947–1948, and then to seventy-five in 1948–1949. Research expenditures went up with the number of projects. The plant devoted RM 3.4 million to research in 1946–1947, RM 4.3 million in 1947–1948, and DM 6.77 million in 1948–1949.[84] During the period from 1946 to 1949, Hoechst scientists and engineers concentrated for the most part on applications of existing knowledge to improve the yield or efficiency of products or processes. Their efforts centered around organic chemistry. As the occupation period progressed, they tended to become more interested in pharmaceuticals and plastics, two of the major growth sectors for the chemical industry in the 1950s. Projects investigating production and product improvement in the textile field and in the area of pesticides and plant protective agents were also well represented in Hoechst research efforts.[85]

Expenditures for research and the number of projects undertaken in the late 1940s appear ludicrously small compared to the later efforts of the Farbwerke Hoechst A.G.: in 1952, for instance, the newly formed successor corporation spent about DM 30 million on research and development.[86] But the earlier figures must be seen in the context of the disastrous production relationships and the restrictions on research between mid-1946 and mid-1948. The expenditures permitted, at the very least, improvement of existing products and processes. In terms of chemical research, though, the occupation period was more related to the late Nazi period than the pre-Nazi or post-1950 eras. Basic research, at the core of German manufacturing success in the chemical industry, languished.

CONCLUDING REMARKS

U.S. policy to break up the Hoechst group into individual factory units and rapid reduction of military government personnel had interrelated effects. Germans obtained a great say in occupation affairs, at least at the operational level; what is more, they had this say as soon as their counterparts in the British zone did. In addition, German *Länder* authorities—especially those in the Hessian Economics Ministry—had the greatest impact of all on policy implementation. The situation was doubly ironic. On the one hand, considering U.S. traditions of strictly limited government intervention, the government's great role in the U.S. zone was remarkable, although the fact that government at the *Land* or local level—and not a higher authority—exercised governmental authority fit in more with U.S. traditions. On the other hand, irony lay in the fact that extensive and increasing power on the part of German authorities in the U.S. zone coincided with growing American power and leadership in western Allied policy in Germany. Once again, though, the apparent contradiction must be qualified: German power remained constrained by the very real, very close, and often strict supervision of overworked U.S. military government personnel. Occupation authorities followed American political tradition by assuming a watchdog function in the economy instead of directly controlling it.

More than anything else, severity and the fostering of an active role of German state government separated U.S. policy from British policy in the early occupation period. In the British zone, business continued as usual, or as usual as was possible during the occupation.

3
BUSINESS AS USUAL: THE BRITISH AND THE I.G. NIEDERRHEINGRUPPE 1945–1946

Great Britain emerged from World War II but a shadow of its former self. The *Wehrmacht* had not occupied its cities, but the *Luftwaffe* had nonetheless inflicted heavy bombing damage on several of them. To pay for the conflict, Britain liquidated many of its overseas assets. After 1945, the British empire steadily eroded, in part because of the war. But this is hindsight. The nation had much to celebrate at war's end. It had stood firm against the Axis powers, for a time practically alone. Its armies played a major role in the African and European campaigns; consequently, Britain was in a strong position to have a say in the postwar settlement. The British empire stood intact for the time being, and by means of agreements reached within the European Advisory Commission and ratified by the Great Powers, the country took over the most industrialized and populous area of Germany.[1]

The British zone of occupation consisted of the northwest section of the old Reich, and included the current-day provinces of Schleswig-Holstein, North Rhine Westphalia, Lower Saxony, and the city-state of Hamburg. In the core of the zone lay the economic heart of the German empire. A relatively small area, the triangle from Cologne to Duisburg and over to Dortmund bounds an area that was (and remains) heavily populated, well endowed with coal (Germany's sole abundant natural resource), and extensively industrialized. Much of the country's iron and steel production and coal supplies emanated from this region. Featuring numerous factories in the Cologne–Leverkusen area and in the Ruhr district, the chemical industry was no exception to the generally high industrial concentration in the area.

Although their zone was blessed with industry, population, and resources, British occupation authorities faced serious difficulties. Because of spatial concentration, destruction to industry in the zone was often more severe than it was in other zones. Transportation and communication outages exacerbated the problem. Most seriously of all, however, the British zone lacked sufficient foodstuffs. With its population swollen with displaced persons and refugees, the zone became the most extreme example of the dilemma of Germany in the immediate postwar period. An area rich in human and industrial potential and in one very important resource, coal, it was unable to feed itself, nor in the short run to export. I.G. factories in the British zone, since they were central to the zonal economy, played a key role in the solution to such problems.

To some extent, the British anticipated the difficulties that would face them in their zone, and thus planned for the occupation. But unpredictable extrazonal forces raised difficult questions: Would Britain impose recently undertaken domestic reforms such as nationalization of key industries on the defeated country? Would the government use its control of Germany's industrial heartland to enhance British industry's competitive abilities? How, in the changed international economic climate and in the face of growing financial difficulties, would Britain be able to afford the spoils of victory in its occupation zone?[2]

PLANNING AND INITIAL IMPLEMENTATION OF BRITISH OCCUPATION POLICY

Unlike the French and to a greater extent than even the Americans, the British were exceptionally deliberate and timely in their planning for the occupation. In March 1944, well over one year before the war ended, the government set up its primary planning organization under the leadership of the Foreign Office. The Economic and Industrial Planning Staff (EIPS) originally had just two formal members, J. M. Troutbeck of the Foreign Office and R. M. C. (Mark) Turner of the Ministry of Economic Warfare, but it later included representatives from all governmental agencies with a stake in deciding the future of the German economy.[3] The government entrusted EIPS with two main tasks: to ensure that various concerned departments of the British bureaucracy had the chance to participate in developing policy toward postwar Germany; and, to coordinate and oversee production of draft directives to military government on the German economy and industry.[4] The EIPS organization thus had a similar function to IPCOG (the Informal Policy Committee on Germany) on the American side, except that it never sent a general policy directive along the lines of JCS 1067 (see chapter 2) to the British military governor. Instead, several different working parties under the auspices of EIPS produced proposals for specific industries. Their proposals in turn served as the bases of British occupation policy.

Circulated in draft form in June 1944 and approved in July, the "Report of the Working Party on the German Chemical Industry" provided the basis for British policy toward the chemical industry. It began by stressing the importance of the chemical industry to the German economy and underlined the need to forego hard and fast recommendations until occupying authorities knew more about wartime developments in, and damage to, the industry. The report then went on to investigate various industry production sectors and ended with a listing of policy proposals covering four major areas. First, the British would eliminate that part of the industry built primarily for war purposes. Slated for elimination were all chemical warfare plants and most explosives factories. Since, however, even the most peaceful economy has some need for explosives (e.g., for mining and roadbuilding), some of the factories could continue to produce, albeit under strict British control. Second, the working party proposed that production facilities employing high-pressure processes—including facilities for the production of synthetic ammonia and synthetic methanol—be dismantled or destroyed. Third, the report suggested that military government personnel investigate and give policy advice on other areas of production including, for instance, dyestuffs and chlorine. Finally, synthetic rubber capacity would remain intact.[5]

Except for the provision allowing for the retention of synthetic rubber production capacity, original British policy toward the chemical industry in Germany bore a strong resemblance to parallel U.S. policy (although the Americans developed theirs independently and somewhat later). Both foresaw elimination of industrial plants related directly (and sometimes indirectly) to the conduct of war and strict control over remaining factories. Both were, furthermore, general policy directives; they postponed detailed policy formation on the chemical industry until the beginning of the occupation itself.

The governmental departments that had participated in EIPS, represented by many of the same officials, set up the British organization responsible for implementing economic policy in Germany. They also appointed its personnel. Organization of the Control Commission for Germany (British Element) (CCG [BE]), Economics Division, began early as well. In July 1944, representatives of the Ministries of Supply, Fuel and Power, Food, of the Foreign Office, of the Treasury, and of the Board of Trade met to consider a draft organizational chart for the Economics Division prepared by Mark Turner. The group agreed which governmental agencies would select branch chiefs for the division, and set October 1, 1944, as the target date for its establishment. In October, a skeleton Economics Division began to function; by May 1945, the Trade and Industry Section (later an independent division) of the Economics Division comprised fifteen separate branches with 600 employees. The main headquarters of the group followed the Advance Headquarters of CCG(BE) to Germany in July 1945. By the end of 1945, Advance HQ was located permanently in Berlin, with the Main HQ situated in Minden in the British zone.[6]

Sir Percy Mills, the chief of the Economics Division, had the responsibility for selecting most of its high-ranking personnel, subject to the approval of concerned governmental agencies. The Board of Trade put slight pressure upon Mills to choose as many civil servants as possible to serve as heads of the division's functional branches. He agreed to try, but soon reported that most British government agencies were unwilling to release senior bureaucrats for service in Germany. Private individuals, including many industrialists, therefore received the call to serve in the higher positions.[7] The men chosen to run the chemical branch of British military government illustrated the tendency well. C. S. Robinson, appointed chief of the branch, worked with Mills and his staff in the Advance HQ in Berlin. At the time of his appointment in January 1945, Robinson worked in the Ministry of Supply. Prior to the war, however, he had been chairman of Imperial Chemical Industries

(General Chemicals) Ltd. and, in this capacity, a member of the board of directors of Britain's largest chemical combine, ICI.[8] M. Zvegintzov, in charge of the Branch in Main HQ in Minden, left the job the August 1946 to return to Shell, the British–Dutch oil giant. The first I.G. Farben controller for the British zone, H. L. Broadbent, was appointed in November 1945 after the British decided to join the other Allies in seizing the properties of the I.G. He was a former employee of Dunlop, the large British rubber and tire manufacturer.

Men with industrial connections thus figured prominently in the control of the chemical industry in the British zone. But what were their relations to one another while they carried out their mission? What was the profile of the staff working under them? Figure 4 shows a schematic diagram of the organizational relationships for control of the chemical industry. The head of the Chemical Branch and his staff acted as part of the Advance HQ in Berlin. Their responsibilities included general policy implementation, supervision of the lower ranks within the branch, and direct contact with the other Allies. The Chemical Branch in the Main HQ at Minden (designated MG ECON 8) was more concerned with day-to-day control of the industry and its provision with raw materials and energy. Beginning its work in December 1945, the British production control apparatus extended from the state through the regional level, and eventually to the *Kreis* (district) level in some parts of the zone. As the unit of control approached the local level, military government personnel had closer contacts with individual German chemical companies. British I.G. Farben Control had a separate organizational structure from the Chemical Branch and was concerned directly with property and control questions relating to the breakup of the I.G. Farbenindustrie A.G.[9] In the months after November 1945, the British zone controller was often in Berlin to meet with the other Allied I.G. Farben controllers, but most members of his staff were located around former I.G. factories. He eventually established primary residence in Leverkusen, the site of the largest British zone I.G. installation.

Zonal chemical control authorities were located close by one another, and some personnel had multiple official positions; this ensured considerable coordination of policy in spite of organizational separation. Hence, chemical branch personnel from both the

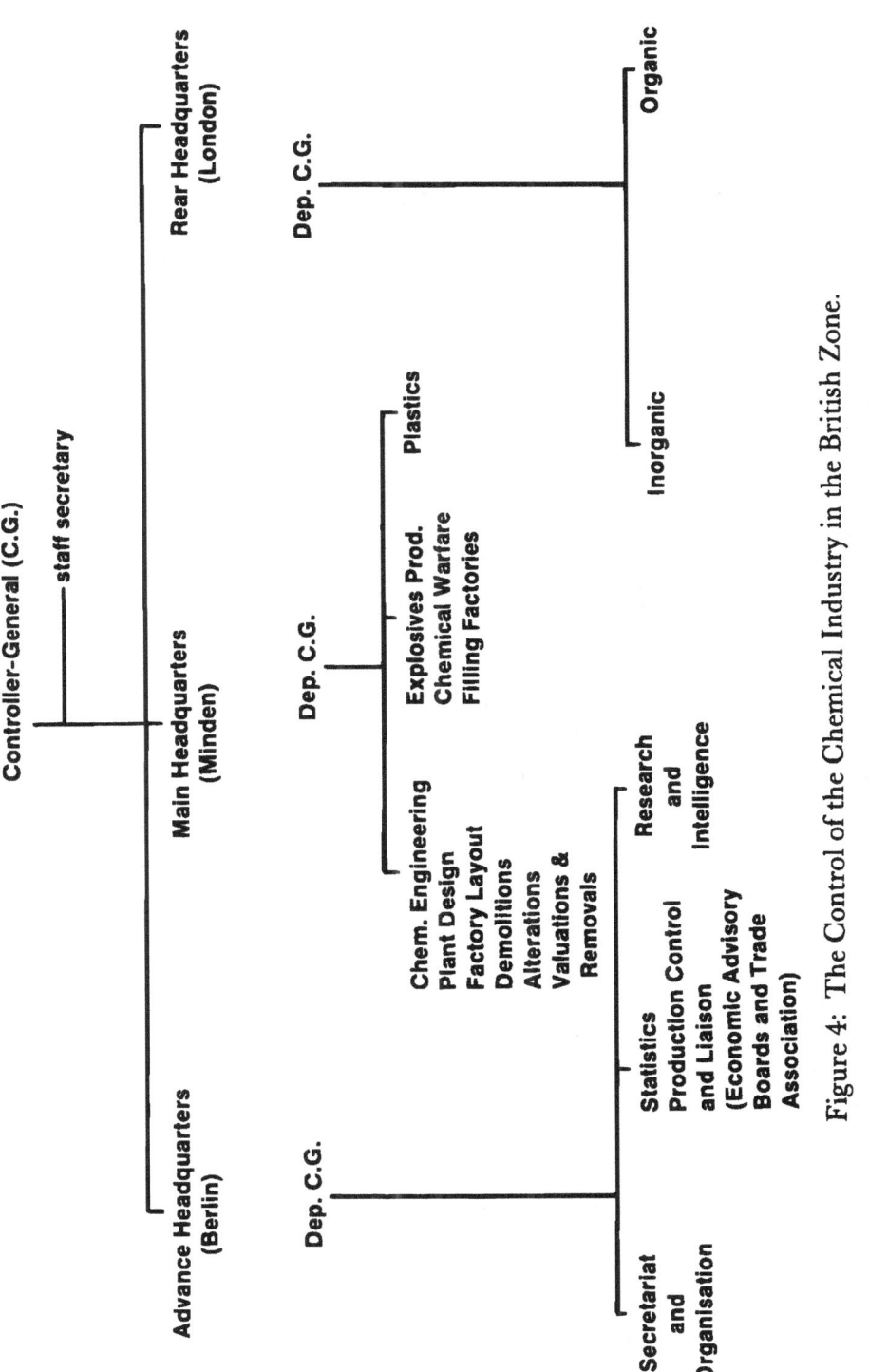

Figure 4: The Control of the Chemical Industry in the British Zone.

Source: Proposal (War Office) 79/Mob/7011. January 8, 1946. PRO FO936/58.

Main HQ and from the North Rhine Westphalia region worked side-by-side with I.G. control personnel in Leverkusen. Several members of the North Rhine Westphalia Chemical Section acted also as zonal controllers for specific product areas.

Multiple responsibilities and limited availability of personnel records cause difficulties in estimating the numbers of persons involved in British control and direction of the German chemical manufacturing. Consequently, study of the problem of connection of chemical branch personnel with British industry is also problematic. Some information exists on the size of the branch. In November 1944, government planners estimated that it would be about forty strong, but the branch grew swiftly in the following year. Requests from British occupation authorities for personnel to staff it increased even more rapidly. By mid-November 1945, the establishment of the chemical branch comprised some 274 administrative officers (of whom 103 were in the headquarters and 171 in the provinces) and 171 staff personnel (88 HQ, and 83 provinces)! During the following two and one-half years, however, the size decreased quickly. By 1948, the French estimated the British staff responsible for running the German chemical industry at just 23 in all (plus, one must assume, German staff personnel).[10]

THE CHEMICAL INDUSTRY AND THE I.G. IN THE BRITISH ZONE OF OCCUPATION

Of the four zones of occupation in postwar Germany, the largest concentration of the chemical industry was in the British zone. In 1936, the area was responsible for about 35 percent of Germany's net production worth in the chemical and chemical–technical industries (excluding asbestos and rubber products). Second in size, with about 28 percent of total production, was the Soviet zone of occupation, followed by the American zone (20 percent), the French zone (11.2 percent), and Berlin (6.4 percent).[11] There were three major focal points for the industry, the Cologne–Leverkusen area, the Ruhr district, and Hamburg. With the exception of the part of the industry located in Hamburg, the heaviest concentration of chemical manufacturing the British zone thus centered in what became the province of North Rhine Westphalia.

The British zone also contained a significant segment of I.G.

Farbenindustrie A.G., but less than the area's share of the German chemical industry as a whole.[12] Heavy investment by the I.G. in middle and eastern Germany from the late 1920s through the war years accounted for some of the discrepancy. The most important integrated set of I.G. production units in the zone was the *Niederrheingruppe*, or Lower Rhine Group, which comprised the main works complex at Leverkusen and its affiliates at Uerdingen, Elberfeld, and Dormagen. The group had close associations with the Duisburger Kupferhütte and a plant at Knapsack, near Cologne.

As is true with most large German chemical companies, the firm that later became the heart of the I.G. Lower Rhine Group was founded in the 1860s.[13] A small chemical dealer in the Ruhr area began to call his business Friedrich Bayer & Co in 1863, and the firm grew steadily, becoming in July 1881 an *Aktiengesellschaft*, or joint stock company. After its incorporation, the Farbenfabriken vorm. Friedr. Bayer & Co. began to invest heavily in research and development, to hire more academically trained chemists, and to draft research chemists into the management of the company.[14]

The 1890s witnessed the beginnings of the transformation of the Farbenfabriken Bayer from a small joint stock company into one of a handful of dominant chemical corporations in Germany—and indeed in the world—by the outbreak of World War I. Bayer's management purchased a new site for the company's central works complex in late 1891. Formerly the property of the small alizarin dye factory of Dr. C. Leverkus & Söhne, the new grounds afforded Bayer an optimal location: they were ten kilometers north of Cologne, sat directly on the Rhine, and presented few physical obstructions to the expansion of a fast-growing industry. Purchase of the Leverkusen site gave Bayer room to expand; the rise to prominence of Dr. Carl Duisberg at about the same time allowed the firm to take advantage of its recent acquisition.

Duisberg, whose career is described briefly in chapter 1, set his stamp on the company from the early 1890s into the Weimar period and beyond. His 1891 memo on the development of the Leverkusen site, when expanded to a detailed plan, became the long-term strategy for development of the firm's new property. It led to a comprehensive concept for utilizing space on the new grounds: it took into consideration present and future transportation needs, access to sewerage and other services, and interdependence of chemical products and processes. The plan represented an insightful forecast of the probable course of expansion of an ever-changing industry.[15] What is more, Duisberg's memos of 1904 and 1915 on the future of the German chemical industry served as a model for and an impetus toward the creation of I.G. Farbenindustrie A.G. in 1925.

Between 1925 and his death in 1935, Duisberg's influence as architect of the concern and chairman of its supervisory board helped build up the reputation and improve the place of the former Bayer factories within the new company. The Bayer group was highly regarded for other reasons as well. Observers inside and outside the industry considered it a forward-looking segment of the I.G. Its reputation for progressive thinking was related in part to the well-planned factory property at Leverkusen. The Bayer group's strong research tradition was yet another basis for its strong reputation and led, after 1925, to a concentration of some of the I.G.'s advanced research laboratories in Leverkusen. Its research facilities included, for instance, the I.G.'s Central Synthetic Rubber Laboratory. In addition, the group's commercial prowess remained unequaled, and its main trademark, the Bayer cross, was probably the most widely recognized trade name for I.G. products. As a consequence, virtually all pharmaceuticals made within the I.G. bore the Bayer symbol, and Leverkusen handled pharmaceutical marketing for the entire trust.

The production and geographic integration of the former Bayer factories and their relative independence from other I.G. works groups constituted another characteristic of the group. In 1936, for instance, 25 percent of Leverkusen's production was utilized either within that factory or by one of the other major components of the Lower Rhine Group (i.e., Uerdingen, Dormagen, and Elberfeld). Two-thirds of its production went to firms and individuals outside the I.G., and only 7.3 percent of total Leverkusen production went to other divisions within the I.G. Supplies from other I.G. divisions, furthermore, made up just 15.5 percent of Leverkusen's net production value, while deliveries from the smaller factories within the Lower Rhine comprised 9.3 percent of Leverkusen's production.[16] The Bayer-related factories thus depended upon one another, but together formed a fairly independent and complete production unit. Leverkusen, the main plant, produced a wide array of organic and inorganic chemicals, many of which served as preliminary products for the other factories in the group. Elberfeld specialized in medicines and plant-protective agents; Dormagen was know primarily for synthetic fibers; and Uerdingen replicated on a smaller scale the broad production of the main plant.

The geographic coherence of the Bayer-associated factories was no less striking. All important Lower Rhine Group production units were in what would later become the British zone of occupation, literally within miles of one another. This was in marked contrast to the Upper Rhine Group of the I.G. centered around BASF. The two central factories of the BASF group were located close together, in Ludwigshafen and Oppau, but BASF had large production units in all four of the later zones of occupation and controlled the I.G. factory at Auschwitz (located in modern-day Poland).[17]

The group's characteristics, Duisberg's strong personality, and his role in the formation of the chemical trust meant that the Leverkusen-based production unit occupied a relatively strong position within the I.G. Throughout the existence of the chemical combine, however, investment in the plants of the Lower Rhine Group compared favorably to that in other western zone plants (although in general I.G. plants in central and eastern Germany were favored for investment over the western zone plants throughout the Nazi period).[18]

During the war, the main plants of the I.G. Lower Rhine Group were relatively unimpeded as they produced goods for the German war economy. Owing to their position within easy striking range of Germany's western border, they were occasionally subjected to air attacks. Damage proved, however, surprisingly slight. Allied air strikes hit the Leverkusen area an estimated thirty-four times between June 1944 and March 1945. Twenty of them were actually fallout from raids directed at Cologne and other neighboring areas. Of the fourteen intended for Leverkusen, only seven were heavy air ones; of the seven, only a few bombs caused significant damage to the plant complex.[19] The war's final skirmishes probably did more damage to the Leverkusen plant than did bombing. In March 1945, American troops stood directly across the Rhine from Leverkusen. Unable at first to cross except with difficulty, U.S. troops shelled the plant at each sign of activity. When the troops arrived to take over the factory on April 14, 1945, however, virtually no one resisted. By April 15, U.S. army forces occupied the entire city.[20]

Life in the Leverkusen works retained an aspect of continuity, and perhaps normality, even during the bombardment and the U.S. takeover of the factory. Although production shut down before American troops arrived, Leverkusen's managers and workers successfully kept the plant's electrical-generating equipment in operation since electricity produced by the factory supplied workers' housing and the town as well. Power generation proved difficult not because of shortages of coal, but rather because the Americans, believing that the Germans were trying to produce war supplies until the last minute, bombarded the facilities every time smoke appeared. Bayer employees overcame the problem by funnelling exhaust into unused production buildings; smoke dissipated slowly and imperceptibly through broken windows and holes in the roof.[21] Lack of long-term disruption in power delivery was paralleled by administrative continuity: the minutes of a regular meeting of the Technical Direction Conference in the factory on April 15 reported:

> During the meeting a delegation of American soldiers appeared under the leadership of an officer. He announced that a curfew would be in force except during the periods from 8–10 and 15–17 hours, and that all firearms were to be turned in. The collection point was to be factory gate 1, and for the residents of private housing [*Eigenheime*] at Dr. Wenk's [office].
> Next meeting: Tomorrow, Monday, 16 April 1945, at 8:30 in K17.[22]

These instances of organizational continuity provide clues about grass-roots resistance to Hitler's schemes for the self-destruction of German industry.[23] In the case of the Leverkusen plant, many of its employees remained committed to their company, or at least to their workplace, and did what they could to protect what was left of the plant from further destruction. Their possession of firearms (alluded to in the American order) may have indicated a willingness to defend the plant not only from Allied attacks but also from agents acting on Hitler's directive. They—their ranks now swelled by returning soldiers—pitched in immediately after the war to clear rubble, repair machinery, and resume production.[24] Whether the Reich's leadership played a role in resisting the Führer's orders for self-destruction or not, it is clear that Bayer workers resisted—

perhaps unconsciously—by virtue of the desire to preserve their workplace.

U.S. troops occupied the plant until June 12, 1945, when they handed formal control over to British forces.[25] The Americans supervised compilation of the first damage estimates at Leverkusen and elsewhere, and the results are telling. Francis Curtis and Mayor Fogler, two investigators for the Combined Intelligence Objectives Subcommittee, interviewed several directors of the Lower Rhine Group at Leverkusen on April 29, 1945. The managers reckoned total damage to the Leverkusen complex to be in the neighborhood of 25 to 30 percent of total property value. Touring the plant themselves after the interview, though, Curtis and Fogler reported that they saw considerable "superficial damage" but little that could not be repaired quickly. They noted that "it is doubtful ... if damage will run to 25 to 30% as stated by the Directors interviewed. Our opinion is that actual damage to process equipment, taking the plant as a whole, is 5 to 10%. Building damage ... is somewhat higher." Their report also went on to point out that much of the plant appeared to be heavily used and some of it outdated.[26] Later Bayer estimates of war damages to the plant tended to confirm the analysis of the Allied investigators. The company's own officials reckoned the extent of damage at war's end was about "15% for buildings and equipment in the factory, not including living quarters." By the time the Bayer report appeared in August 1946, "54% of the extent of the ... destruction" had been repaired or reconstructed.[27]

Other major plants of the group were also damaged only slightly, as table 4 indicates.

British occupation officials thus assumed control over a relatively lightly damaged set of I.G. chemical plants on June 12, 1945. Their full contingent of military-government personnel arrived in Germany shortly thereafter, equipped with long-prepared and much-discussed policy instructions.

INITIAL POLICY IN ITS DOMESTIC AND INTERNATIONAL CONTEXT

The British did not implement their policy on the postwar German chemical industry and on the I.G. in isolation; rather, external pressures constrained policy choices. Key areas of external policy

Table 4. War Damage to Major I.G. Lower Rhine Group Factories

Plant	Number of Employees	War Damage (in percents)
Leverkusen	13,500	15
Dormagen	3,500	1
Uerdingen	2,500	22
Wuppertal-Elberfeld	1,250	none

Source: Herbert A. Broadbent. "Report to the Committee of Control Officers as to Seizure of Properties of I.G. Farbenindustrie AG in the British Zone." April 5, 1946. Appendix. PRO FO 1039/644.

pressure and constraint included: the domestic political situation in postwar Britain; the commercial interests of the British chemical industry and attempts by industrialists to influence postwar policy; and the international political context of the postwar period. It will be useful to survey each area briefly.

When the Potsdam Conference reconvened after the British general election of July 1945, the American and Soviet leaders faced a new negotiating partner. Instead of the Conservative prime minister of the wartime coalition British government, Winston Churchill, Soviet leader Josef Stalin and American president Harry S Truman now negotiated with the new head of a Labour government, Clement Attlee. Labourite Ernest Bevin replaced former foreign secretary Anthony Eden. On the domestic scene, nationalization of key industries and corporations (e.g., the Bank of England in 1946 and the National Coal Board in 1947) and other related measures (e.g., the creation of the National Health Service in 1948) followed the change in government. At the same time, many of the wartime controls over the British economy remained in force, and British planning agencies continued to function. Rationing of some essential commodities continued until 1954 and sometimes became even more severe than during the war itself. Foreign exchange transactions were also closely monitored.[28]

Apparently, the Labour victory in mid-1945 affected policy toward Germany primarily by causing it to veer in the direction of socialization of key German industries.²⁹ Given the links between the ruling Labour Party and the German SPD and given the Labour government's domestic actions, the change is unsurprising. The actual defeat of British-zone socialization measures in 1948 and its causes lie outside the scope of this study. Several additional aspects of the impact of British domestic developments on their occupation policy are worth considering, however. For one thing, Germany was removed from the British political scene both in terms of physical distance and in terms of its political environment: in other words, German occupation politics took place in a different, *international* arena separate from British politics. Policy in Germany was therefore bound to be different from that at home, even in the crucial area of socialization, since in Germany British authorities had to consider the wishes not just of British voters, but also those of the other occupiers and the Germans. What is more, many of the same civil servants carried out British policy toward Germany (and indeed important aspects of British domestic and foreign policy) after the Labour victory as had done so before. The result was a drag on radical change in policy.³⁰ Since there was greater participation in British military government from societal elite groups that were often even more conservative than the civil servants (e.g., the military and industrialists), the inertia was probably stronger in Germany than at home. In sum, the impact of British domestic policy on the occupation was significant but slight, and often indirect. Its impact on the chemical industry was minimal.

A key interest group on the British economic and political scene, the British chemical industry itself, had a much greater effect on policy. I have already mentioned the industrial connections of personnel employed in the Military Government Chemical Sections in Germany. Clearly, these men had ample opportunity to recast or to form relationships with their German counterparts, to conduct negotiations on marketing or patents with them, or to try to stifle future German competitive ability. They probably made use of the opportunity, although there is no evidence for this.³¹ Documentation on other areas of attempted industry influence is more readily available: the areas include the exploitation of German industrial know-how for British industry and attempts by industrialists to prevent possible resurgence of German competition.

British exploitation efforts parallelled those of the Americans (see chapter 2). Government-backed investigators "collected" and detained German scientific and technical personnel for interrogation and possible use in Allied projects, especially in defense research. Of much greater interest to industrialists was the second prong of the program: there was a massive effort to collect and disseminate information on German products, processes, and patents developed during the war. It constituted what was probably the largest transfer of technology across national borders of all time. Representatives of industry influenced the effort substantially, since the British government asked industry to help set up the program. Industrialists suggested targets for investigations of German industry.[32] Industry also supplied technical experts to visit the defeated nation to carry out the studies.[33]

After 1945, the investigations gradually slowed down as policy turned increasingly toward rebuilding the German economy: reconstruction, after all, implied some attempt to provide protection for German industrial secrets. British and American groups officially discontinued inquiries by the end of June 1947 in the Bizone, the combined British and U.S. zone.[34] Although industry continued from time to time to try to wrest information from military government on the capabilities of German competitors, British authorities became increasingly unreceptive to industry pleas. A request from Imperial Chemical Industries (ICI) in 1948, for instance, asked for records relating to costing of dyestuffs and dyestuff intermediates from the former I.G. companies. The government turned down the request with the advice "that if ICI want the information for which they have applied they can obtain it by process of free negotiations with the German officials" of the old I.G.[35] Government bureaucrats thus began to resist continued industry efforts to gain information on its once and future competitors.

British industrialists also attempted to influence their country's policy directly in order to limit the competitive ability of their German counterparts. Many historians accept without question that businessmen were successful in this endeavor. Indeed, a common interpretation is that desire to curb future German competition characterized British economic policy in Germany.[36] The reality is, however, more complex.

Fairly soon after the beginning of the occupation, representatives of the British chemical industry were well aware that their German counterparts would soon pose a renewed threat to British markets.[37] The Chemical Branch asked ICI in early 1947 to grant a license to a German chemical firm to manufacture one of its (ICI's) patented products, hexachlorocyclohexane, an insecticide. ICI's reply was unequivocal:

> It is in our view certain that in a few years' time Germany will once again be a serious competitor in the world markets and we do not wish to assist them in establishing the manufacture of new chemicals such as this insecticide. . . . We are consequently not agreeable to the proposal that we should allow this company to manufacture H.C.H. and grant a license.[38]

British chemical industrialists also recognized that the interest of military government in increasing exports from the British zone would conflict with their own interests. At a meeting in February 1946 between government personnel and representatives of ICI to discuss proposed production levels for the German dyestuffs industry, Sir Frederick Bain of ICI summarized the dilemma well. He maintained that he was troubled by:

> a. The probability that the Control Commission would positively encourage the exports of German Dyestuffs in order to alleviate demands on the British taxpayer to finance imports; and
> b. The danger that Germany would in that case regain her leading position in the world market for dyestuffs and chemicals.[39]

Because they were disturbed by the specter of renewed German competition and by apparent conflicts of interest with government objectives, representatives of the British chemical industry asked for a say in policymaking with a view toward restricting their German counterparts. In early 1946, a representative of the British rayon industry, S. Courtauld of Courtauld's Ltd., requested that the industry be consulted in policymaking for the level of the German rayon industry.[40] Shortly thereafter the director of the British Rayon Federation submitted a paper to the Board of Trade (BOT) outlining the Federation's policy proposals on "The Post-War German Rayon Industry." German industry, the report recommended, should be strictly controlled; Germany should not be allowed to use lower wage rates to obtain price advantages in world markets; exports from the occupied country were not to exceed prewar levels and were furthermore to be restricted to former markets; finally, British industry was to be consulted in future policy alterations.[41] The Association of British Chemical Manufacturers (ABCM) initiated similar action for the chemical industry in general, and for the dyestuffs sector in particular.[42]

Such efforts of British industrial interest groups on their own behalf were neither unexpected nor unusual: similar groups and individuals from French and U.S. industry took similar actions. Reception of the recommendations and proposals by government and MG personnel responsible for the formation and implementation of industrial policy was much more significant. The reply of Stafford Cripps, the head of the BOT, to the letter from S. Courtauld in early 1946 was representative of governmental response. Cripps welcomed industry suggestions on policy but gave little hope for their implementation. He stressed that government–industry efforts to collect German technical information, combined with the wartime destruction to German industry, would afford British industry a period in which to gain ground in the world market.[43] Cripps and other officials believed it was up to British industry to make good use of its relatively favorable position vis-à-vis German industry. The head of the Trade and Industry Division of CCG(BE), Eric Seal, put the sentiment well:

> The situation surely is that the British and other Allied producers have a splendid opportunity whilst Germany is prostrate of building up a strong competitive position. They ought to be told to concentrate on building up this competitive position and not to rely upon the elimination of competition by a perpetuation of control. . . . To my mind, it is regrettable that the present trend of thought in British industry appears so often to favour smothering competition, at the expense of the consumer and the taxpayer also if need be, regardless of the fact that it is only by competition that pre-eminence in industry can be built up. It strikes me as a defensive, and indeed defeatist, state of mind.[44]

On the other hand, in the initial occupation period British government policymakers agreed without reservation to the idea of industry participation in policy talks. At a meeting between BOT and British Rayon Federation representatives in July 1946, the government agreed to keep industry informed and up-to-date on German industrial policy.[45] The Control Office for German and

Austria in London agreed to consultations with the Federation of British Industries in 1946 as well.[46] In addition, the government permitted representatives of British industrial trade associations—including the British Rayon Federation and the Dyestuffs Mission of the ABCM—to visit Germany to study and to report policy recommendations on German industry in the British zone.[47]

British policymakers became less and less sympathetic with industrial input into the formation of policy as time went on. The trend parallelled that in exploitation of German scientific and technical information by British industry. Initial government sympathy with industry interests produced permission for the British Rayon Federation and the Dyestuffs Mission to go to Germany. By 1947, it became a promise reluctantly fulfilled. A letter from Derek Wood (BOT) to R. Keeling (FO, German Section) depicted the change:

> If you will refer to the papers on the rather confused subject of consultation with industry, you will see that both you and we are committed to letting these people [the Dyestuffs Mission] go out [to Germany]. While developments in Germany in the last year have greatly reduced the ground for consultation of the type which the industry had in view, visits such as the proposed are nevertheless of considerable value to Industry here which must needs make increasing allowance for the revival of German competition. Examination by competent experts on the ground and in conjunction with the Germans themselves produces a much better picture of the probable future trends than any amount of correspondence through inexpert channels. Moreover, discussions with the competent authorities in Germany enables [sic] the British side to blow off steam. They deem this to be their prerogative because of their contribution, as taxpayers, to the German economy. In these respects the visit of the Rayon Federation was an outstanding success, if measured only in the falling off of memoranda submitted to our respective Ministers.[48]

British policy between September 1946 and November 1947 therefore changed from concentration on the desires of British industry to attention to the needs of the British zone and its expense to the British taxpayer. The costs of running the occupied area became a key policy determinant by early 1947.

Although domestic political developments and industry influence played a part in the early postwar period, the impact of international affairs gradually grew in prominence.[49] The "international political context" for postwar British policy toward defeated Germany consisted of a number of fundamental, and to some degree interrelated, sets of developments. For one thing, in the immediate postwar period, the British faced a series of crises in their currency, the pound sterling, which were the direct result of the financing of the British war effort. Growing antagonism between the Soviet Union and the western powers was another important determinant of postwar British occupation policy. Finally, the United States gradually assumed the initiative in postwar western Allied occupation policy. Increasing American influence in German affairs had an especially pronounced effect on British policy, in particular after the economic fusion of the British and American zones into the Bizone on January 1, 1947.[50]

Taken together, the three elements had a significant impact. Financial weakness forced the British to try to encourage exports from their zone to pay for the enormous costs of occupying it; their subsidy to the zone was approximately 80 million pounds during 1946, most of which had to be paid out in scarce dollar reserves.[51] The shortage of funds also forced the British to rely more and more on U.S. financial support. Increasing tensions with the Soviets and the growing influence of the United States in occupation affairs had the effect of turning British policy toward encouraging thoroughgoing economic reconstruction and export recovery in the Bizone, and later in all three western zones; the policy shift involved gradual increases in production limits for German chemical firms as well as encouragement of export. The growing importance of the United States was reflected in the perceived necessity to follow— at least in form—U.S. policy with regard to I.G. Farben. In mid-November 1945, when the British finally joined the other occupation powers in seizing the assets of the I.G. and installing a zonal controller for the former I.G. factories, they consciously— albeit reluctantly—followed the U.S. model of seizure.[52]

At the same time, this is not to say that British policy became subject completely to developments on the international scene, or that British policy consisted solely of slavishly imitating the Americans. Zonal policy on the chemical industry and the I.G. developed to a great extent within the zone itself, in particular owing to the growing influence of German bureaucrats and industrialists.

THE BRITISH ZONE'S CHEMICAL INDUSTRY AND OCCUPATION POLICY

In spite of the noticeable effects of extrazonal forces, there were elements of independence and stability in on-site British policy implementation. This was especially true for the plants of the former Lower Rhine Group of the I.G. Farbenindustrie. One of the key British actions in their treatment of the former I.G. factories was to do nothing to break up the interdependence among them: from the beginning, the occupiers allowed the Lower Rhine Group to remain intact. What is more, although the fate of Dormagen remained uncertain into the 1950s and although Uerdingen, too, was slated to be removed from the group through 1950, the British side never really questioned that the core factories of Leverkusen and Elberfeld should continue as a single production unit regardless of how the I.G. itself was to be broken up.[53] Another crucial instance of British policy independence and of their quest for stability was the retention of Dr. Ulrich Haberland as the works director of Leverkusen and as director of the Lower Rhine Group. Haberland was the only manager of a major I.G. works group who had been in his position during the Nazi period, and who retained it without interruption following the German surrender. For this reason, it will be necessary to return to his case shortly.

Maintenance of the Lower Rhine production group and retention of Haberland were both cause and effect of generally friendly relations between British occupation personnel in Leverkusen and their German subjects. What is more, British controllers for the most part allowed German managers to run the firm as they (the managers) wished. Occupation officials concentrated instead on strict control of war factories (e.g., explosives manufacture) and looser general supervision of other production areas.[54] A natural consequence of the policy was a strong element of continuity in technical and commercial relationships within the production group from the 1920s (and indeed from the earliest years of the twentieth century) through the occupation period. A further consequence was a certain degree of independence, of prominence, and thus of influence upon industry developments on the part of British-zone chemical industrialists. Foremost among them was the works director of I.G. Farben Leverkusen, Dr. Ulrich Haberland.

Haberland was still a relatively young man at the end of the war.[55] Born in Saxony in 1900, he served a brief stint in the army in 1918 before studying chemistry at the University of Halle. Receiving his doctorate in 1924, he continued to work in the university as an assistant until 1925, when he entered private industry. Changing employers in 1928 proved to be a decisive step for the young chemist. At the beginning of that year, he took a position in the Uerdingen works of the Lower Rhine Group of the I.G. Haberland rose rapidly through the ranks, becoming director of the Inorganic and Organic Intermediates Section at Uerdingen in 1931, and head clerk (*Prokurist*) at the factory in 1935. At only thirty-seven years of age, Haberland became director of the Uerdingen works.

The war advanced Haberland's career to new heights. In 1943, he was appointed director of the Leverkusen works of the I.G. and shortly thereafter took over the direction of the four factories at Leverkusen, Elberfeld, Dormagen, and Uerdingen as head of the Lower Rhine Works Group. The appointment signified the immense amount of trust and responsibility afforded the relatively young chemist and manager. But it was important for another reason as well: in appointing Haberland to the post, the I.G. direction chose not to include him in the company's managing board. Appointment to the board was normally a matter of course for the director of the Lower Rhine Group, but the omission in Haberland's case was apparently quite deliberate. The I.G. direction purposely kept Haberland off the board in the hope—later justified—that he would not be removed from his position as group leader in the case of a German defeat. He could thus serve as an element of continuity and as a point of concentration for the reconstruction of the firm.[56]

Indeed, although the I.G. was never reformed, Haberland's tenure as head of the "Bayer Group" during the occupation period proved instrumental in determining the shape of the postwar western German chemical industry. While keeping his group intact with British support, Haberland acted as a key adviser both to military government and to the later federal government on questions related to the chemical industry, and especially on the breakup of the I.G.[57] He also attracted talented former I.G. executives to the Lower Rhine Group from other zones and other works groups. A key instance of this, of major consequence for the Bayer group in the 1950s, was the exodus to Uerdingen of a group of chemists and engineers from the massive Leuna Works in the Russian zone of occupation. The group specialized in manufacturing caprolactum, the most important intermediate for production of the synthetic fiber Perlon.[58] But perhaps the most famous example was Dr. Karl Winnacker, later chairman of the managing board of Hoechst A.G., for whom Haberland secured a position at Knapsack when U.S. authorities prohibited Winnacker from working in their zone.[59] Many former I.G. managers and researchers from the Russian zone of occupation came to the British zone to work under Haberland. His prior position as a prominent I.G. manager made him highly visible to those unwilling or unable to work in other zones, while his eye qualified him to recognize talent for the Bayer firm.[60]

Throughout the occupation period, Haberland enjoyed excellent relations with his British overseers. The British allowed his group to act as a unit from the beginning. They encouraged him in his efforts at reconstruction. They permitted employment of such talented persons as Winnacker and the Leuna caprolactum group. When Haberland was arrested by U.S. authorities in 1947 and transported for trial to Nuremberg, Haberland "was released as the result of an application by the British Authorities, who pointed out an irregularity in his arrest in that he had never been a member of the pre-war Farben Board"[61]—despite the fact that Haberland's high-level managerial responsibilities had made him a "board member" in all but name during wartime.

Under Haberland, the Lower Rhine Group began the long process of recovery and resurgence after 1945. Before examining that process in more detail, however, a closer look at the remaining other western zone of occupation will be necessary. The next move is south, to the French zone of occupation and the former I.G. Upper Rhine Group.

4
THE TECHNICAL LIMITS TO EXPLOITATION: THE FRENCH AND THE BADISCHE (BASF) 1945-1948

French-occupied Germany—composed of the Saarland, South Baden, Württemburg-Hohenzollern, and Rhineland Pfalz—had the smallest population of the four zones and only limited natural resources. Aside from the Saar area (which was in any case treated separately) and the chemical industry in the area around Ludwigshafen on the Rhine, the zone's industrial base was relatively weak as well.[1] Developments in the zone were nonetheless crucial to the postwar economic history of western Germany. For one thing, they are compelling since, in the initial occupation period, French policies apparently differed sharply from those of the British and Americans. More importantly, French actions in Germany were bound up with the origins of postwar Franco-German *rapprochement*, which has been a key determinant of European political and economic development since 1945. Economic historian Alan Milward has argued that "it is the economic objectives of French European policy [in the postwar period] which were the true determinants of a more lasting western European settlement."[2] Germany, of course, figured prominently in the determination and realization of French objectives. Throughout the period 1945 to 1948, German and French technical personnel played a pivotal role in determining the pace and direction of developments in Franco-German relations.

THE CHEMICAL INDUSTRY IN THE FRENCH ZONE

Although there were chemical factories in each of the provinces in the French zone of occupation, the character of the industry was different in each case. South Baden's chemical industry featured several large and modern plants;[3] that of Württemberg had, "on the whole, an artisanal character";[4] Rhineland Pfalz (more specificially, the area immediately around the city of Ludwigshafen) contained the bulk of the industry in the French zone. In all, factories in Ludwigshafen and vicinity manufactured an estimated 80 percent of the area's chemicals. One plant complex in the area towered above all the others in importance: in terms of number of employees, "the single factory of the I.G. Farben (currently known as BASF) represents by itself a little less than three-fourths of the chemical industry in the Z.F.O. [Zone Française d'Occupation]."[5] What is more, BASF produced about 50 percent of the turnover of the entire zone's chemical industry and constituted the French-occupied area's largest industrial unit.[6] To a much greater degree than in the other western zones, a single I.G. works group dominated chemical manufacturing; BASF epitomized the chemical industry in the Franch zone.

Badische Anilin- und Soda-Fabrik is one of Germany's oldest and most successful industrial corporations. Founded in 1865 by Friedrich Engelhorn, the firm took part in the explosive expansion of the German area chemical industry in the 1860s. Engelhorn's idea was to establish a facility to produce chemical intermediates for his five-year-old dyestuffs factory in Mannheim. The then tiny village of Ludwigshafen, directly across the Rhine from Mannheim, served as the site of the new facility. Since the village placed relatively few spatial or legal limitations on its growth, the Ludwigshafen factory was able to expand rapidly, and the firm came to be synonymous with the name of the town itself. Even today, with Ludwigshafen a city of about 160,000 inhabitants, BASF occupies a dominant position in its municipal life.[7]

Artificial dyes formed the basis for the firm's growth in the latter half of the nineteenth century. After massive expenditures for research and development, it became the first to synthesize indigo dye on a large scale. Success in this field provided the funds to erect a second, even larger BASF plant in the initial years of this century. BASF managers chose Oppau, about one mile north of Ludwigshafen, as the site. The new factory was to manufacture yet another new product: in 1913, utilizing innovative high-pressure and high-temperature technology, Oppau began production of synthetic nitrogen for fertilizers and explosives. Demand for synthetic nitrogen's products grew dramatically during and after World War I, and the firm expanded production in other areas, too.

By the 1920s, the two main BASF factories and their associated plants manufactured a full range of chemical products, including dyestuffs, pharmaceuticals, fertilizers, and other products of the new synthesis technologies. The factory complexes at Ludwigshafen and Oppau alone manufactured more than 2000 different products. Ludwigshafen specialized in heavy chemicals, intermediates, dyestuffs, and other organic chemicals. Oppau gained prominence for high-pressure chemistry, and alone provided over one-third of total German fixed nitrogen during World War II. In terms of production technology, too, the factories differed:

> At Ludwigshafen, many operations were carried out in small scale, batch equipment, such as is used in many kinds of chemical manufacturing, and a large number of the buildings were old, comparatively fragile, and very susceptible to bomb damage.... Oppau [in constrast] can properly be considered the prototype of the plants at Leuna, Heydebreck, and elsewhere, which produced fixed nitrogen, organic chemicals, and synthetic fuels by processes involving the use of hydrogen under high pressure.[8]

Together, though, the two main production units of the Upper Rhine Group were essentially independent from the other I.G. works groups (and from the BASF-associated factories to the east). Both, however, relied heavily on coal, coke, and coal gas from the Ruhr and the Saarland for raw materials and energy.

From the beginning of the company's existence, its management possessed a distinctive style: BASF managers tended at times to take risks with seemingly reckless abandon; they always took pains to be, or become, well connected in German politics; finally, BASF's directors have been innovative in terms of products, techniques of production, and organization. Until the Nazi period, BASF could be counted among the most risk-prone corporations in the world. Competition to be the first to produce synthetic indigo in the late nineteenth century was extremely bitter and intense. BASF, under the leadership of its chairman of the managing board, Heinrich von Brunck (himself a talented chemist), invested much of its capital and over seventeen years of concerted effort before it succeeded in 1897 in large-scale production of synthetic indigo. After 1897, the strategy paid off handsomely for the corporation, but it could have caused very heavy losses indeed had it been unsuccessful. The story went on: the young chemist Carl Bosch headed BASF's pioneering effort to develop synthetic ammonia in the years just before the outbreak of World War I. With government aid and wartime demand, the investment paid off once again. In the 1920s and 1930s, the firm again invested heavily—in a scheme that at times more resembled gambling than corporate risk-taking—in the development of synthetic gasoline and rubber.[9]

Political astuteness on the part of BASF managers and their ability to be useful in difficult times often allowed the firm to weather the crises brought about by their *Risikofreude* (enthusiasm for risk-taking). Enormous expansion of BASF's nitrogen fixation operations occurred during War War I with the financial support of the German state. The firm thus accelerated the transition of a new technology from the pilot plant to the mass production stage by convincing state bureaucrats to subsidize the process.[10] In return, the state received an invaluable service: had it not been for the nitrogen fixation facilities of BASF, fertilizer and ammunition production in Germany would have come to a halt in 1915 thus bringing the German war effort to an end. Aid came to the company (by that time, the center of the new I.G. Farbenindustrie A.G.) from late Weimar regimes in the form of duties on imported gasoline. That and more direct aid from the Nazi regime for the firm's synthetic oil and rubber production plans rescued BASF and the I.G. from severe financial difficulties.[11] In the Third Reich, the actual placing of company personnel in government offices superceded mere good relations with the state. I.G. officials participated in production and investment planning, which contributed to the firm's massive expansion in the Nazi years.[12] Perhaps the near disaster brought on by synthetic fuel development induced a new element of caution in firm policy; perhaps caution accompanied the gradual aging of the concern and its leadership. In any case, political activity became a partial substitute for entrepreneurship by the Nazi period.

Nevertheless, especially in its early years, BASF was an innovative corporation. It pioneered, for example, in the application of science to industry, a key characteristic of the new industries of the late nineteenth century, and particularly of the chemical and electrical industries. Heavy investment in synthetic indigo, nitrogen, gasoline, and rubber development and the influence of the trained chemists von Brunck and Bosch at the company's highest levels were but the most prominent examples of this tendency. Organizational innovation parallelled product and process innovation at BASF. The firm was one of the first to have a technically trained customer service staff, for example, although organizational innovation occurred on a larger scale as well. Bayer's Carl Duisberg first conceived of and helped implement the consolidation of Germany's major chemical firms into a conglomerate to cut down on competition at home and to strengthen the world position of the industry; BASF's Carl Bosch was a central figure in the actual incorporation of the companies on December 9, 1925, into a single trust, the I.G. Farbenindustrie A.G. Since all of the other corporate components sold their shares to the Ludwigshafen company, Farben was in fact the legal successor to BASF. Bosch became its first chairman of the managing board. BASF was its heart in terms of productive capacity, technical achievement, research, and management talent.

When intensive bombing raids on German industry began in late 1944, Allied strategic bombing experts assigned a high priority to striking synthetic fuel and rubber factories. The BASF plants at Ludwigshafen and Oppau, partly because of their advanced synthesis technologies and spatial concentration, and partly because of their geographic proximity to Allied air bases, received a disproportionate share of bombing damage. BASF's two main facilities were indeed by far the most heavily damaged of the major western German I.G. plant complexes. In all, the two plants were subjected to sixty-five bombing attacks. Ninety-four percent of 1500 factory buildings received some damage. Production capacity at the plants sank to about 40 percent of peak level. Total losses were estimated at more than RM 400 million.[13]

French military government inherited this heavily damaged plant complex at war's end in 1945. Given the extent of the damage, the tone of a French report from July 1946 is astonishing, since it claimed that, "with the I.G. Farben installations at Ludwigshafen, France has received one of the jewels of the German chemical industry."[14] But then again, the French had very good reasons for being happy with their inheritance. Damage to buildings on the factory sites proved not so serious as to prevent resumption of some production. Plant capacity out of commission at war's end—about 60 percent altogether—could be reduced substantially by simple and inexpensive repairs. German machinery and know-how stood at the disposal of the French, and the most important element in the BASF success story—its scientific and managerial staff—was largely intact.

In fact, it is difficult to overestimate the importance of BASF to the French military government, and indeed to France itself. The zone's largest single industrial unit, BASF, compared favorably in size with France's own most important chemical concerns. In 1949, the company amounted in size to about 15 percent of the entire French chemical industry (in terms of productive capacity, personnel employed, and turnover). If only large firms were considered, the proportion jumped to 20 percent. In terms of production of several important chemicals, the comparison was even more impressive: in 1949, BASF manufactured by itself half again as much methanol as did the entire French chemical industry, over half as much ammonia, and 40 percent as much fertilizer. It produced, in addition, one-fourth of the dyestuffs and one-fifth the chlorine in comparison with the entire French chemical industry. Not a single French chemical firm was anywhere near the size of BASF. In the words of a French report,

> By comparison to *other chemical corporations* in Europe or in America, BASF occupies an honorable place which it will improve upon in 1950. . . . And in a general fashion one may say that it has the same importance as Montecatini, Ciba [-Geigy], or Leverkusen among the European corporations. It would be necessary in contrast to put several French corporations together in order to constitute an ensemble of BASF's importance.[15]

INITIAL FRENCH POLICY TOWARD THE I.G.

Having experienced three invasions by the Germans in less than seventy years and having lived under German ocupation for most of the Second World War, the French were clear in their perception and evaluation of the German question. In August 1944, a typical

French analysis proclaimed that "*The German problem* is, *for France*, the essential problem. . . ." The author went on to point out that a single principle would guide the solution to that problem: "To assure the security of France for numerous generations, in other words to eliminate the danger from Germany permanently, is thus *the premier task*, a sacred task, which faces the French state."[16]

Initial French policy included a series of complementary diplomatic, political, military, and economic measures designed to safeguard France and Europe against future German aggression. It pushed toward decentralization of political and economic authority and favored dismemberment of Germany into smaller political units. The Ruhr area, the heart of German industrial might, was to be placed under international control, and the policy foresaw demilitarization of the defeated country. France and its wartime Allies would keep Germany in check through defensive alliances.[17]

Economic measures to control future German aggression were fundamental to the French program. As the August 1944 study pointed out, "The most virulent militarism would be inoffensive enough if it did not have the formidable *materiel* indispensable to modern armies, and which only a powerful industry is in the position to furnish. . . ." Underlining the centrality of economic control of Germany to attaining French objectives, the author maintained that "if, at the moment when the total occupation of the Reich ceases, Germany retains self-determination with regard to and the unity of its industry, a third world war may be considered as inevitable."[18] Thus, German industrial might should be weakened through various measures, including dismantling of excessive capacity, prohibition and/or control of production of key items, and deconcentration of the economy through breaking up large industrial units and banks. The similarity of these aims to initial American and British objectives is evident.

Underlying French economic policy was the realization of the central role of the chemical industry in the German war effort in both world wars:

> . . . after having allowed Germany during the war from 1914–1918 to continue the manufacture of explosives by bringing synthetic nitric acid into production, it [the German chemical industry] has permitted the preparation and the realization of the current war by implementing

massive production programs for synthetic fuels, synthetic rubber, explosives, plastics, replacement products, and new products of all types. The final sensational weapon developed by the German, the V-2, is a chemical weapon. The motor and the explosive are chemical.[19]

In the French view, the attempt to fulfil this enormous role during the wars had led to extreme concentration in the German industry with the result that, by 1945, two or three large firms, "among them the firm I.G. [Farben], . . . control more or less directly probably 80 percent of the German chemical industry." Concentration in turn had allowed the German chemical industry to play a double role, on the one hand serving as the basis for much of Germany's military might, and on the other acting as a "power for economic and political expansion" in its own right.[20]

Such considerations dictated specific elements of the French program for the industry. For one thing, French officials sought to severely curtail foreign trade in chemical products. Since the industry was heavily dependent upon export trade even during the relatively autarkic Nazi period, and since it required several key imported raw materials to function normally, the effect of this policy would be to throttle its productive capability. Second, French military government in Germany wanted to control allocations of basic materials to chemical firms, including, above all, coal. German chemical production was to be held in check through destruction of, or reparations (in the form of physical plant and primary or intermediate materials) from industry's "excess capacity." Just what would constitute excess capacity remained unclear. In any case, the program of destruction and reparations would achieve the twofold goal of trimming the power of the German chemical industry and of helping its counterparts in Allied countries—and in France in particular—back onto their feet.[21]

The French had special interest in this final point. Severe restrictions on German industry would have the effect of leaving open important export markets in Europe and overseas and would hamper future attempts by German firms to reenter them. French officials were anxious, with the aid of German plant, materials, and know-how, to fill the export gaps and encouraged domestic producers to take full advantage of the situation just as did British authorities:

It rests with French industrialists—through their individual efforts, through adoption of cooperative solutions, which will without doubt be necessary, and with the help of appropriate measures taken by public authorities—to realize the necessary conditions for this increased and renewed regime of production, and in particular to arrive at production costs which permit them to withstand international competition in order to attain and retain the place in the world to which they [French industrialists] have a claim.[22]

BASF in particular would serve the French economy in various ways: with scientific and technical information; with production for France; and with commercial contracts favorable to French firms.[23]

On the surface, the main thrust of French policy toward the German chemical industry was straightforward: French policymakers would use any and all means at their disposal to gain their main objective, which was to curb forever the possibility of German aggression. Because a major policy component was a series of economic measures designed to undermine the warmaking (and also the peacetime economic and political) potential of German industry, the program signified France's clear intention frankly to exploit German industry in their zone in its own service. The chemical industry was a prime target for this policy of exploitation.

On a deeper level, though, French policy was ambivalent.[24] In order to exploit fully the resources of its zone and to achieve other policy goals, French military government could not pursue a program of unadulterated rapacity. German production in service of the French economy could be achieved on the scale needed only if at least some German technicians and managers resumed their positions and made some repairs to their factories. Unmitigated implementation of many policy elements, in other words, would have had the effect of killing the proverbial goose that laid the golden egg. Consequently, the French avoided sheer exploitation of German industry. The ambiguity and complexity of what they actually did in Germany became clear in the course of implementing policy at BASF.

FRENCH CONTROL OF GERMAN CHEMISTRY

In comparison to the other powers that occupied defeated Germany in late-spring 1945, the French were ill-prepared to administer their zone of occupation. It is small wonder. The Germans had occupied France from 1940 to 1944. Only after the liberation in 1944—and in fact well into 1945—did the French become certain that they would have an area to administer in Germany.

Since French occupation troops faced extensive difficulties in the immediate postwar period, preparation would have been a distinct advantage. The problems of the I.G. successor in the zone counted among the most severe. Relatively high rates of destruction to the BASF have already been mentioned: its facilities were much more heavily damaged than those of any former I.G. plant complex of similar size and importance.[25] A second major problem French military government officials faced was coal supply. Coal served as the major raw material and energy source for the chemical industry at the time, yet was present in significant quantities only in the Saarland, the Russian zone, and, most importantly, the British zone. Moreover, the coal industry was even more plagued with problems than other industries in postwar Germany. Finally, the chemical industry in the French zone was dependent upon markets abroad and in the other zones of occupation, and upon imports for supply of certain key raw materials. Relative isolation of the four zones of occupation from one another in the immediate postwar period and the almost total exclusion of German industry from export markets thus constituted serious problems.

Lack of preparation for the occupation and the severity of the problems faced in the initial period did not impede thorough fulfillment of many French policy aims. Chemical industry production levels in the French zone, for instance, far exceeded those of the Bizone at least until mid-1948: by the currency reform (June 1948), zonal chemical manufacturing had reached 91 percent of its prewar level even as bizonal chemical production languished at just over 50 percent of its prewar rate. Figures for two important BASF products, unfortunately only available from January 1947, show this development in greater detail (see table 5).[26] French and German technicians achieved these levels of production, furthermore, through a policy under which relatively little was dismantled: as a German bizonal administrator observed in 1947 regarding proposed dismantlings from BASF, "From developments to this point, one may assume that the French military government leans rather toward seizing production from the works than toward an actual dismantling of the installation."[27]

Table 5. Production of Two Key Products at Ludwigshafen and Oppau, 1947–1949 (in metric tons)

Year/Month	Primary Nitrogen	Methanol (raw)
1947 Jan.	3091	472
Feb.	2984	1277
Mar.	3360	494
Apr.	4744	1405
May	6679	1725
Jun.	6950	2137
Jul.	6450	2526
Aug.	7700	3296
Sep.	6930 (6163)[a]	2107 (462)
Oct.	6010	1163
Nov.	6166	2500
Dec.	6822	721
1948 Jan.	6304	1281
Feb.	5250	1750
Mar.	6400	2250
Apr.	6100	2000
May	7750	1450
Jun.	7800	1700
Jul.	7800	2800
Aug.	8500	2500
Sep.	n.a.	n.a.
Oct.	n.a.	n.a.
Nov.	n.a.	n.a.
Dec.	n.a.	n.a.
1949 Jan.[b]	9520	1041
Feb.	8867	2091
Mar.	8800	1825
Apr.	9000	2186
May	7591	1121
Jun.	6993	n.a.
Jul.	9121	1648
Aug.	9163	2362
Sep.	9025	990
Oct.	10014	1117
Nov.	11676	3526
Dec.	14607	2688

[a] Column figure from Report of September 1947; figure in parentheses from Report of November 1947.

[b] Figures for 1949 for calendar months. Thus, number of days varies.

Source: Compiled from *Bulletin Mensuel*, I/1947–III/1948; BASF, "Productions," n.d., MGFr, Caisse 1840, Dossier 43; BASF, "Etats de produits fabriqués," MGFr, Caisse 1840, Dossier 32.

Relatively high production levels had, from the French point of view, a desirable result: sales of chemical manufactures were channeled into directions useful to their own interests. BASF, for example, sold most of its production in 1947 and 1948 either within the French occupied area to help cover zonal needs (it imported relatively small quantities of chemical products),[28] to France itself to aid in fulfillment of production plans there, or to the other (mostly western) zones of Germany to pay for coal imports or imports of key raw materials.[29] A good example of the latter was a generally functional but frequently problem-ridden exchange program from 1946 to 1948: the Hoechst plant in Frankfurt-Hoechst in the U.S. zone received nitrogen from Oppau for further processing into fertilizer; in return, U.S. military government guaranteed shipment of a certain amount of coal to the French zone to be used largely for the production of chemicals.[30]

Occupation authorities also promoted the interests of the French chemical industry. One method was to invite science and engineering students from universities in France to Ludwigshafen and Oppau to observe processing techniques and hardware available at BASF. More importantly still, military government promoted a series of contracts between BASF and French chemical firms; unsurprisingly, they were usually to the advantage of the latter. In 1948, for instance, Francolor obtained—in large part through the good graces of French occupation officials—exclusive rights to market many BASF products outside of the German territories.[31] Considerable international uproar over this particular deal resulted in alteration of its terms. Nonetheless, a report from the French director general of BASF from October 1949 listed a new series of contracts concluded in the past year that were advantageous to French chemical firms.[32] Finally, the Chemical Subdivision of French military government in Baden-Baden maintained an office in Paris during the occupation period. The Paris bureau served as both a clearinghouse for requests from French companies for information and materials from Germany and as a liaison between military government agencies and the Chemical Section of the French government Ministry of Industrial Production.[33]

Why, despite limited planning, were the French so effective in inducing such stunning production levels at BASF? One reason was the relative compactness of the French zone itself, and, more to the point, the geographical and production concentration which BASF in Ludwigshafen and Oppau represented. After all, within a few kilometers along the left bank of the Rhine in this single plant complex stood over half of the French zone's chemical industry (in terms of turnover), and a good 10 percent of the total German chemical industry. The plants of the Bayer group in the British zone, taken together, may have rivaled the main BASF plants in size, but the Bayer works group included other plants more or less far afield, in Uerdingen, Dormagen, and Elberfeld. In addition, Bayer represented a much smaller percentage of the total British-zone chemical industry. True, the industry in that zone was concentrated as well, but to a much lesser degree than in the French zone. The U.S. zone's chemical industry was in many ways the structural opposite of its French zonal counterpart: there was, and is, a concentration of the industry in the Rhine–Main area, stretching along the Main from Frankfurt to the Rhine and north to Wiesbaden. But even this grouping could not be considered an overwhelming concentration of production facilities, since they were scattered throughout the zone's provinces. The BASF plants were thus simply easier to control than those of the other former I.G. works groups.

Spatial concentration of the objects of control was parallelled by a relatively centralized organizational structure. Since it was heavily staffed with chemical authorities, the French administration assumed an extensive and detailed supervision of the chemical industry—and especially of the BASF—that was qualitatively different from that in the other western zones. Throughout the occupation period, central French authorities responsible for the chemical industry were located in Baden-Baden as a subdivision of the military government division, "Production Industrielle," which consisted of a director and his staff, along with several branches responsible for the control of specific product areas. Through 1946, the branches were more numerous; in June 1947, however, they were consolidated into three major ones, the Branche Minérale, the Branche Organique, and the Branche Parachimique. Two nonproduction branches supplemented the three, one responsible for "Planning and Allocations" and the other responsible for "Reparations, Restitution, and Transportation."[34]

Unlike the other western zones, where occupation authorities responsible for the chemical industry at the provincial (*Länder*) level were fairly important, parallel officials in the French zone were relatively uninfluential. In early 1947, for instance, a military government commission pointed out that the activity of the former I.G. Farben factories "is not at the level of the Land, or even of the zone, but represents rather a national or even international interest. What is more, the orientation of their activities is currently complementary to the French chemical industry. . ." Thus, planning and allocation for the former I.G. would occur at the central level, and provincial authorities were informed of the decisions reached there.[35] The French military governor himself, General Pierre Koenig, endorsed this procedure once again a year later.[36]

Authorities responsible for the zone's I.G. plants were more significant than provincial authorities. They worked in tandem with central administrators to regulate key chemical production facilities at Ludwigshafen and Oppau. French specialists occupied key positions in each factory, including those of director general, technical director, and director of maintenance. They worked directly on a day-to-day basis with German personnel to keep the BASF factories going in spite of coal shortages and equipment breakdowns.[37]

To carry out the control effort, the French employed a large contingent of administrators and technicians. The numbers involved are telling. In 1948, all of the western military governments reduced their staffs drastically. Nonetheless, the number of French personnel involved in the control of German chemistry was, at approximately fifty, more than half again as large as the British chemical control staff. Those responsible for the industry in the U.S. zone amounted to just three, or one-tenth the British contingent. In spite of heavy cutbacks later in 1948, the number of French chemical control officials, at twenty, remained on about a par with the British staff. The Americans retained qualified personnel only at the highest levels.[38]

French administrators also held their positions for a relatively long time. J. P. Fouchier was chief of the central chemical section (although the title changed from time to time) from 1946 until 1950. Many of his top assistants remained on the job throughout most of the occupation period, and personnel employed directly at

BASF plants by French military government also tended to remain at their posts for most of the same period.[39] In contrast, British personnel were generally in their positions for a shorter time, and the American staff, small as it was, changed remarkably frequently.

The result of concentration of production, an extensive supervisory staff, and relatively long-term tenure for military government personnel was a level of control unseen in any of the other western zones, or in any of the other former I.G. factories. Records of the French authorities at BASF, for instance, included detailed monthly reports on the production of individual products in prose and in statistics, as well as more general reports on the state of the plant, on allocation of key raw materials, and on other items of general interest, all submitted by French control officers. Reports of meetings between French technical authorities and their German counterparts in each of the two major BASF factories demonstrated that Franco-German teams discussed in detail repair schedules, solutions to production bottlenecks, and even sales strategies.[40]

French and German technicians thus cooperated to solve common problems, and this cooperation constituted the third and final factor in explaining the effectiveness of the French in bringing about high levels of production at BASF. The very idea of *cooperation* between the French and German *Erbfeinde* (hereditary enemies), especially in the early occupation period, goes against the grain of all expectations. For one thing, the apparent acquiescence of BASF technicians and workers in the fulfillment of French aims is striking: the French were, after all, seizing German production for their own needs, deciding export quotas, and exploiting their position as occupiers to impose agreements on BASF favorable to their own industry. The definitive reasons for the seeming lack of resistance, owing to incomplete documentation,[41] must remain in the realm of speculation. Probably, though, the extent and intensity of French supervision played a role. Until 1948, even passive sabotage of French orders by BASF technicians and managers must have been impossible: a French officer stood over the shoulder of every major BASF staff member. In addition, bureaucratic inertia probably was significant in dampening possible resistance to French policy. BASF personnel had carried out their orders under the Nazis, and they continued to carry them out under the occupation administration. As long as those "orders" involved maintaining and improving production at Ludwigshafen and Oppau—by all accounts a difficult and time-consuming task—BASF personnel could probably find little to oppose in French policy.

The final probable reason for lack of German resistance to French programs was the behavior of French supervisory personnel; for, ironically, tight French control of BASF in pursuit of their own policy aims coincided to a remarkable extent with the interests of German technicians and managers. Nowhere was the irony clearer than in the area of the maintenance of BASF plants. The French encountered the technical limits to exploitation in this vital area, as the catastrophe at Ludwigshaften in 1948 illustrated.

On July 28, 1948, in the middle of a warm summer afternoon, a series of explosions ripped through the BASF factory in Ludwigshafen. After the fires were extinguished and the smoke cleared, the extent of the damage became apparent: 200 men died, 500 were seriously injured, and another 1500 were lightly injured. A total of 10 percent of the Ludwigshafen complex was completely destroyed, and another 5 percent seriously damaged. In all, destruction to capacity was estimated to be about 15 percent, a figure that was startling in view of the fact that wartime destruction to Ludwigshafen's capacity was about 25 percent.[42] The Oppau plant was, of course, not damaged by the explosion, since it lay downriver.[43] Production in the Ludwigshafen plant resumed the following day at 50 percent of the prewar level (the level of July 27 was 80 percent), but the result of the explosion was nonetheless a severe blow to the chemical industry in the French-zone occupation since repairs would be costly and time-consuming.

The catastrophe was a tragic expression of deep and long-recognized problems at BASF, the problems of worn-out machinery, of often slipshod repairs, and of limited maintenance. Heavy wartime use of existing machinery and the shortage of capital for new equipment accounted above all for the difficulties. The effects of overuse and capital shortage plagued BASF plants throughout the occupation, partly because haphazard delivery of coal and coke involved production slowdowns and even stoppages.[44] Generally such slowdowns and stoppages represented no real danger to the plants, especially if there were warnings that cuts in coal deliveries could be expected. But, when they occurred in winter, plant equipment was often in severe danger. In December 1946, for example, the French technical director at Oppau, R. Scheffer, noted in a report on the successful shutdown of the plant owing to lack of energy supplies, "We have got through this one with luck, but it is a risk it would be better not to take in the future." Everything possible should be done, he added, "so that one does not run the risk in winter of seeing the factory shut down rapidly by frost without there being time to take indispensable measures."[45]

The shortage of routine and preventive maintenance was the most basic and critical difficulty of all. Director of the Maintenance Service at Oppau, A. Taillefer, pointed this out from at least the late summer–early fall of 1947. Taillefer, whose remarks applied in his view to the Ludwigshafen factory as well, kept up a running campaign to change and improve maintenance policy at the BASF plants. He claimed that lack of preventive maintenance represented a grave danger in terms of possible damage to the factory itself, possible loss of human life, and possible long-term production halts. He recommended increasing the volume of maintenance and returning insofar as possible to a routine program of preventive maintenance. In closing, he noted that the realization of his suggestions "will involve, certainly, a monetary augmentation in outlays. But these outlays will end up paying off when their effects have had time to manifest themselves (in a few months or even in a few years)."[46]

Taillefer's repeated warnings and suggestions paid off in a way, if too late to avert the catastrophe. On May 18, 1948, he was named director of a new Sub-Direction of the Technical Services Direction of the French administration of BASF. The office was meant to coordinate and improve planning and implementation of technical services for BASF as a whole. From that point on, maintenance of the two factories stood under a single director, who was also responsible for coordination of materials supply and energy matters for the plants.[47] The explosion at once confirmed Taillefer's worst predictions and eliminated the maintenance work virtually entirely. For the immediate and foreseeable future, repair work would take up most of the time and energy of the maintenance staffs of both BASF plants.[48]

One must be cautious in analyzing the role of the French technical staffs of BASF in development of occupation policy. Nonetheless, it is clear that the French technical personnel at the BASF factories thought in terms of long-term production strategies for which maintenance and careful management of the factories were crucial. These men may have had in mind long-term and direct French dominance of BASF. They may have thought in terms of long-term cooperation between French and German industry. Ultimately, however, their intentions are not important: the central point is that they thought in terms of getting BASF back into production to stay. This involved by implication running the firm as any German manager would have done.

Furthermore, the warnings of the technical staff may have encouraged a more open attitude at least toward joining the French zone with the Bizone, and perhaps active interest. Evidence for this contention is spotty and circumstantial, but also logically connected. BASF was the largest chemical producer in the French zone and indeed the zone's most important industrial unit. Short-term exploitation of the firm was approaching its limits, as the reports from the technical staff indicate. To be able to continue to produce for the sake of France, investments in equipment, repairs, and maintenance at the BASF factories were necessary. France had very little to give at the end of World War II, especially to a German firm. Reorganization of the technical services just before the catastrophic explosion shows a certain awareness at higher levels of the problem. The prospect of joining with the other two western zones into a Trizone would ease the flow of raw materials to BASF and reintegrate to a larger degree the western German chemical industry. BASF's problems would then to some extent be shared by all of western Germany and by the U.S. and British authorities.

CONCLUDING REMARKS

One of the most striking contradictions of French policy in postwar Germany was between the push for decentralization of political and economic life on the one hand, and the actual implementation of centralized action on the other. The two major I.G. Farben factories of Ludwigshafen and Oppau, together the heart of BASF and of the chemical industry in the French zone, could theoretically have been separated organizationally and technically. Instead, BASF was treated as a unit, thus maintaining an important center of the chemical industry in western Germany. French bureaucratic institutions recapitulated this contradiction: much key planning and allocation for the chemical industry was carried out centrally in Baden-Baden.

Confrontation with the realities of the German situation in the occupation period also brought out another element of the dual thrust to French policy. On the one hand, the objective of French military government was to use industry in its zone to supply and to support French industry. French authorities were somewhat successful in this regard. On the other hand, it is possible to look at the export performance of BASF in a very different light. In fact, in 1947 and 1948, almost four-fifths of all of BASF's sales by value were within the three western zones of occupation.[49] Much of the western German chemical industry thus relied upon BASF for supply of basic chemicals for further processing, while BASF needed coal and other supplies from the other zones. By force of existing technological and organizational interconnections, the French permitted and even encouraged what they had wanted in fact to prevent: the technological and organizational reintegration of the German chemical industry. Steps taken in this direction during the period 1945–1948 were slow and halting. Nonetheless, the direction became increasingly apparent as contacts, especially among the three western zones, became more and more routine.

Parallel to the growing reliance on the German market was the increased reliance on German administrators and technicians to run the industry. The French moved in this direction much more slowly than did the Americans, of course, who by 1946 had handed over significant tasks of administration of reconstruction and production to Germans except at the highest levels. Nevertheless, the French, too, had to turn over some responsibility to the Germans following the cut in chemical-industry personnel from fifty to twenty in 1948. A prominent example of the trend was their employment of Dr. Heinz Krekeler as a chief negotiator and liaison with the Bizone. Krekeler, later consul of West Germany in New York, was born in 1907 and graduated with a doctorate in chemisty from the University of Berlin in 1930. In 1934, he joined the I.G. to work as a chemist at Oppau in the field of butadiene synthesis. Krekeler became the works assistant to the manager at the newly erected gasoline synthesis plant of the I.G. at Heydebreck in upper Silesia during the war. It was not his only connection with the chemical firm: his uncle acted as head of the Lower Rhine Production Group of the I.G. centered at Leverkusen from 1926 to 1933 and as such sat on the firm's managing board. His uncle was also chairman of the I.G. Tea-Büro (I.G. technical committee office) from 1925 to 1932, and moved into the supervisory board of the chemical giant in 1933. In 1946, Heinz Krekeler began to represent French interests in what would become the Bizone. The extent and content of correspondence between him and French-zone chemical authorities indicated the trust placed in him.[50] By mid-1948, the French gave increasing powers of negotiation to other German officials as well.[51]

Ambivalent and often contradictory aspects of French policy in Germany abounded: desire to limit industry stood against successful encouragement of production; the stated goal of decentralization of the zonal economy contrasted with the centralization of direction in Baden-Baden, and of production in the Ludwigshafen area; the aim of tying the chemical industry in the zone to its French counterpart clashed with the necessity for increasing ties to German chemical producers in other western zones; French control of the industry in their zone was undermined by the use of German administrators and technicians. In each case, of course, the stated policy aim was at least partially fulfilled. But the existence of the contradictions, the tensions between stated policy and German reality and within the policy itself, represented an area of fundamental ambivalence from at least 1945. Consequently, the French policy changes of 1948—gradual *rapprochement* with the other western zones through French participation in the currency reform of mid-1948, the slow move to Trizonia, and finally allowing the creation of the Federal Republic of Germany—represented less a break with previous policy than a change of emphasis within it.

If, however, contradictions facilitated the restructuring of French policy, what were the causes for the change? The historian of the French zone, F. Roy Willis, in a second, more general book on the subject in 1968, stressed the outcome of the Moscow Foreign Ministers Conference in 1947.[52] The rejection of French policy proposals by the Russians meant that the French were forced more and more into the position of having to make accommodation with the west, and especially with the United States. Others have stressed alternative explanations for policy alterations: the economic and political troubles on the French home front, and/or the increasing tendency to tow the American line in Germany in order to receive U.S. aid.[53] The impact on policy change of technical personnel within the zone might well be added. German and French technicians at BASF pursued the common goal of maximizing the firm's output. The recognition of the limits to French policy by occupation overseers at BASF may have provided added impetus toward reevaluation of French tactics in Germany. Together, the explanations demonstrate the constraints posed by international developments and the effect of French domestic policy upon occupation authorities, while recognizing the significance of internal zone developments.

The French zone joined with the other western zones in enacting a key currency reform in June 1948. From this point, French zonal developments were no longer separate from parallel developments in the other western zones. Even before, in early 1947, the British and Americans agreed to combine their zones. The history of the western German chemical industry as a whole therefore forms the subject of the remaining chapters.

PART III
THE RETURN TO WORLD COMPETITIVENESS: THE I.G. SUCCESSORS, 1947–1951

On January 1, 1947, the British and Americans fused their zones economically. The advent of the Bizone had more than just formal significance. Seen in light of the growing tension between east and west and the increasingly obvious financial desperation of the British, French, *and* Germans, 1947 marked a major turning point in the occupation. Thereafter, policy in the western zones took on an increasingly American cast: American objectives came to dominate western Allied practice in Germany. The following three chapters thus examine chronologically major issues during the period 1947–1951, including the effects of Allied policy, the growing impact of the Germans on the Allies, the reconstruction of German chemical firms, and the resurgence of the West German chemical industry.

5
THE CONTINUING CRISIS
JANUARY 1947–JUNE 1948

In July 1946, U.S. secretary of state James Byrnes and British foreign minister Ernest Bevin agreed to combine the American and British zones of occupation economically into the Bizone. Full political unification was eschewed for the moment, and the French and the Russians were invited to join in the economic fusion. Byrnes and Bevin thus avoided a complete breach with the other Allies. But the formation of the Bizone made it clear that the British and Americans were prepared, if necessary, to go it alone in occupation policy. It also signaled that policy had changed. Limited reconstruction and positive action replaced punishment and passivity as the primary leitmotifs of Anglo-American policy in Germany.

The creation of the Bizone had far-reaching consequences for the chemical industry. British and American occupation officials responsible for the industry began to meet formally to discuss and to coordinate policy. They began to implement programs in common through bizonal control institutions. The result was steadily increasing unity in British and American policy toward the chemical industry. Furthermore, the United States took over an increasing proportion of the costs of administering the Bizone, and its representatives therefore demanded an increasing say in determining the political and economic direction of the combined area. Policy unity thus took on a pronounced American cast.[1]

The founding of the Bizone also involved the creation of a unified, superzonal German economic administration. German bureaucrats in the two zones responsible for the chemical industry were able to coordinate their actions more effectively to attempt to "administer the shortages" of the Bizone more efficiently. Chemical industry personnel from the bizonal economic administration formed the core of the Chemical Section [*Chemiereferat*] of the Federal Ministry of Economics and thus ensured continuity between bizonal policies and procedures and those of the federal period.

Creation of the Bizone did not cause, but was accompanied by, a worsening of the situation of individual chemical factories. Until late 1946, factories had been able to rebuild to some degree. It was often possible to resume production in vital areas by cannibalizing parts from damaged plants and by running down reserves of raw materials and energy. By winter 1946–1947, continued export of coal to the detriment of domestic industrial production, transportation breakdowns, and generally low coal production throughout the Ruhr district meant that nearly exhausted reserves could not be replenished. The harsh winter of 1947 exacerbated the situation still further. The result was that many chemical factories of the former I.G. came to a production standstill. Chemical production stagnated in the winter of 1946–1947, and there was another series of setbacks in the following winter. Throughout the area, by late 1946 the slowdown in recovery necessitated first Allied, and then German, governmental action toward rehabilitation and long-term recovery of the I.G. Farben successors. Those endeavors focused primarily on the scourge of all areas of the economy in the early occupation, coal supply.

CHEMICAL PRODUCTION IN THE BIZONAL AREA, MID-1945 TO MID-1948

Figure 5 shows the development of the bizonal production index by month from January 1946 until June 1948 for all industry and for the chemical industry. The index is based on 1936 production levels in what became the combined area.[2] Clearly, the chemicals index tended throughout to be somewhat higher than that for all industry. Moreover, the chemical production index fluctuated much more than that for all industry. This was largely due to production stops in, and the subsequent reactivation of, large chemical plants throughout 1947.

These trends bear more detailed inspection. Although figures for 1945 are unfortunately not available, it is likely that the index stood near zero in April, May, and into June 1945. Thereafter, it re-

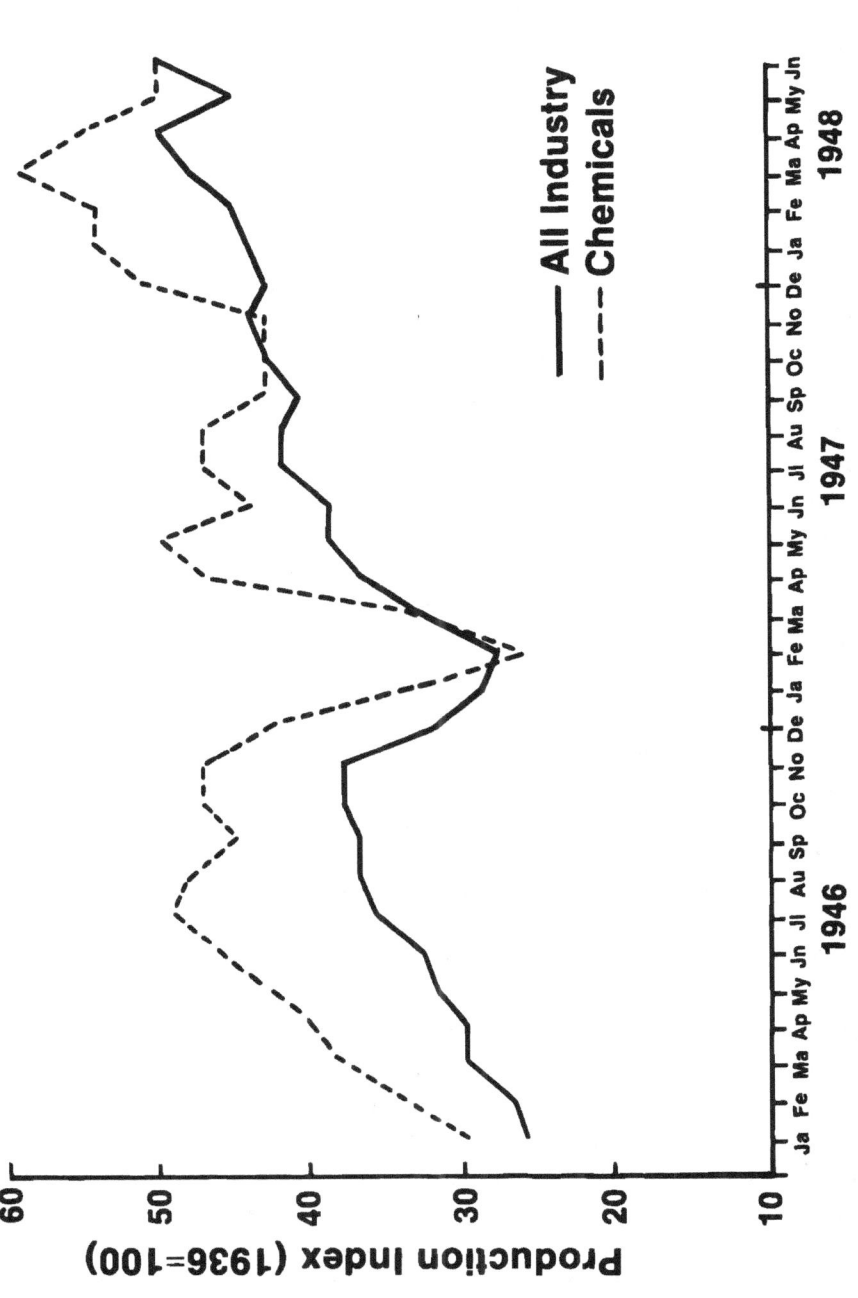

Figure 5: Development of Bizonal Production for all Industry and for Chemicals, January 1946–June 1948 (not seasonally adjusted).

Source: Wirtschaftsstatistik der deutschen Besatzungszonen 1945–1948 (Oberursal: Europa-Archiv, 1948), p. 123.

covered to about 30 percent of the prewar production level. The upward trend continued into the summer of 1946, when the chemical index in the bizonal area reached almost one-half its prewar level. The surge leveled off until near the end of the year. The stasis probably reflected the general malaise of the German economy in late 1946 and early 1947.[3] The country's industrial base was, after all, surprisingly intact after the war and industry was therefore able to recover astonishingly quickly. By mid-1946, however, economic revival itself was beginning to try the limits of the German infrastructure. Systems of transportation and production of energy and raw materials—all among the most hard-hit sectors of the economy owing to overuse and overproduction during the war and to bombing losses—simply began to collapse.

The slowdown would become even more serious when cold weather compounded infrastructural problems. From November to December 1946, the index dropped from 47 to 43 percent of its 1936 level, followed by a disastrous drop to a low of just 26 percent of the 1936 norm in February 1947. Chemical manufacturers reached a point they had not seen since late 1945. Recovery of chemical production proceeded as rapidly as its fall, and the index lay between 43 and 50 percent of the 1936 level from April through November 1947, and was 50 percent or more from December 1947 to June 1948. Production even reached an index level of 59 percent in March 1948. Still, chemical production in the Bizone virtually stagnated between July 1946 and November 1947. This was in contrast to the development of production for industry as a whole, the index for which crept gradually upward for most of the same period.

The bizonal chemical production index conceals important variations, including those by product, by zone, and by firm. Table 6 (1936 = 100) presents the production index from 1946 to mid-1948 for a number of important basic chemicals. The index of production of some key products—including nitrogen, sulfuric acid, and calcium carbide—consistently equaled or exceeded that for the bizonal chemical industry as a whole. What is more, the indices for sulfuric acid and nitrogen did not dip in the winter of 1947, but rather continued their upward trajectory. On the other hand, production of several other important chemical products, most notably soaps and detergents, was consistently lower than the industry average. The reasons for these variations are not clear. Occupation authorities, after all, encouraged production in all of these areas,

Table 6. Production of Important Chemicals in the Bizone, 1936, 1946 to mid-1948 (1936 = 100)

	Nitrogen (fertilizer)	Sulphuric Acid	Calcium Carbide	Soap/Washing Powder	Paint/Varnish/ Laquers	Coal Tar Dyes
1936 m.a.[a]	100	100	100	100	100	100
1946 m.a.	50	34	49	36	n.a.	n.a.
1947 m.a.	63	63	45	33	n.a.	10
1948 Jan.	85	63	63	32	45	19
Feb.	80	63	61	30	51	26
Mar.	89	70	70	29	52	29
Apr.	88	71	71	37	58	29
May.	89	73	71	30	54	23
Jun.	90	70	77	32	52	19

[a] m.a. = monthly average.

Source: Calculated from OMGUS, *Report of the Military Governor, Statistical Annex Number 45* (March 1949), p. 121.

Table 7. Chemical Production in the U.S. Zone of Occupation
(1936 = 100; not seasonally adjusted)

Year/Period	All Industry	Chemicals
1946	37	45
1947	46	41
1946 1st Quarter	28	35
2nd Quarter	34	48
3rd Quarter	42	51
4th Quarter	45	48
1947 1st Quarter	34	26
2nd Quarter	48	49
3rd Quarter	50	47
4th Quarter	50	43
1947 July	52	49
August	51	52
September	48	39
October	50	34
November	51	41
December	49	53
1948 January	50	58
February	54	62

Source: OMGUS, *Report of the Military Governor, Part 23: Statistical Annex Number 32* (February 1948), p. 34.

but performance remained uneven. One reason for relatively high production of basic chemicals may have been relative ease of manufacture: they were produced directly from raw materials in short supply but still available. The laggards—soaps, paints, and dyestuffs—required not only basic chemicals but also the application of still more coal for raw materials and power. Production therefore tended to suffer.[4]

Bizonal production of chemicals prior to the currency reform also varied by zone of occupation. Through 1946, as Table 7 shows, the U.S. zone produced chemicals at levels not significantly at variance with those of the Bizone as a whole (see fig. 5). Zonal production was, however, much more sensitive to the impact of the winter of 1946–1947. By February 1947, chemical manufacturing in the American area lumbered along at just one-fifth its 1936 rate, as compared with a rate of just over one-fourth for the Bizone as a whole. The index recovered as spring arrived, but during the autumn of 1947 chemical production in the zone still stood at about one-third the 1936 rate. The comparable figure for the Bizone as a whole was 43 percent of the 1936 rate of production. Throughout 1947 and into 1948, production of chemicals in the U.S. zone was more volatile, and often very much lower in comparison to the average monthly levels of 1936 than was chemical production in the British zone.

Why was this the case? The chemical industry in the U.S. zone had an important characteristic that made it more susceptible to the problems of 1947 than its British zone counterpart: it was dependent on the British zone for most of its coal. Thus, when the German transportation infrastructure began to break down in 1947, it was the outlying U.S. zone that suffered most. Low coal output and poor transportation meant that there was an increasing tendency for the coal of the British zone to be used there rather than transported to other zones. The result was systematic underdelivery of already insufficient allocations of coal to the zone. The Provincial Economics Office (*Landeswirtschaftsamt*, or LWA) Hessen complained in June 1947, for instance, that

> the Hessian chemical industry has been consistently under delivered in the past. In the month of May, too, the largest and most important coal consumer in Hesse, the Farbwerke Hoechst, received only 1900 tons of its quota of 9300 tons (reckoned on the basis of hard coal).[5]

Complaints continued into 1948, as the LWA claimed that "the supply of coal to the chemical industry for January and February 1948 is to be described as extremely short, since only 35–40 percent of the allocated quantities of coal for January were delivered."[6] The chief of the Mining and Metals Section of the Industry Branch of OMG Hessen, Lt. Col. Henning B. Dieter, was ironic in his evaluation of the situation:

> It may be that . . . uneven allocations [of coal] will lead to a general recovery, but the issue seems doubtful at Land level. We recall that in May [1947] the tremendous allocation to the NR-Westfalen chemical industry was rationalized on the plea of a Buna and soap program.

Presently, our tire industry is well supplied with coal, but reports that Buna is about to run out. To compensate for this, we also have no soap.[7]

The final major set of variations within the chemical production index was by firm. Of primary concern here are the I.G. successors in each zone, the Bayer complex in the British zone and the Hoechst plant in the U.S. zone. Unfortunately, although the firms undoubtedly reported production levels of various products to zonal and bizonal statistical authorities, those numbers have not survived in German state, Allied, or firm archives. Those statistics that do exist are not directly comparable for the two major former I.G. works in the Bizone; therefore, some indirection will be necessary. Nonetheless, some conclusions may be drawn from figures for the intermediate chemicals production section of the Hoechst factory and for one important organic chemical produced at Leverkusen.[8] Between the disastrous second quarter of 1945 (when production levels were close to zero) and the third quarter of 1946, Hoechst turned out chemical intermediates at an increasingly rapid rate. Output declined markedly in the last quarter of 1946 and the first of the next year. Recovery ensued in the second quarter of 1947 and continued through the second quarter of 1948. The trend thus followed that of development of production in the U.S. zone through 1946 and diverged slightly—in the continuing improvement from the second quarter of 1947—in 1947 and early 1948. But the figures conceal one important thing. Most production in the Hoechst intermediates section for the period was accounted for by the output of nitrogen and fertilizers. Changing production rates for these chemicals caused most of the variation through mid-1948. Other product lines in the area (e.g., inorganic chemicals, building materials, and technical gases) featured slow but steady—and virtually unbroken—growth from the second quarter 1945 through mid-1948.[9] In spite of coal shortages, worn-out equipment, and all of the other problems of the early occupation period, Hoechst was able to produce with steady improvement throughout the entire period from mid-1945 to mid-1948. Similar trends are in evidence for other Hoechst sections (e.g., dyes and pharmaceuticals) as well.[10]

Development of production in Bayer Leverkusen's organic section presented a contrasting case, at least on the basis of information on one key product, methanol. Leverkusen's production engineers were unable to put methanol into production from January 1945 through February 1947. (From December 1946 through February 1947, the entire factory shut down owing to shortages of coal, which accounts for some of the downtime.) Manufacture resumed in March 1947 at just a trickle, 20 tons, drastically lower than the 1937 monthly average of 540 tons or the peak production average of 1430 tons per month in 1941. In the following month, however, methanol output jumped to 670 tons. For the remainder of the year and into 1948, it varied enormously from month to month from a low of 170 tons in August 1947 to a high of 1130 tons in October.[11]

Meager quantitative information on production at Hoechst and Bayer in the immediate postwar years makes it necessary to use caution in generalizations. Additional, qualitative information is, however, available about the situation at both factories. Leverkusen, owing to coal shortages, shut down completely for three months in winter 1946–1947 and again for the month of June 1947 despite its proximity to the coal fields. In contrast, the Hoechst factory's distance from coal producers made its shortages even more severe; yet the concern did not shut down. Production development at each factory from mid-1945 to mid-1948 was therefore exactly the opposite of what one would expect on the basis of chemical production trends in the factories' respective zones. Shutdowns at Leverkusen were a sign of the plant's very real problems with coal supply, but were also a sign of the relative independence and strength of the firm's leadership. For they were *not* the result of running out of coal: rather, the plant shut down to build up stockpiles and, from all indications, to try to force German and Allied authorities to increase allocations and deliveries to the plant.[12] In contrast, the Hoechst factory continued to produce, albeit often at low levels, regardless of the coal situation, an indication that its representatives were less independent than their counterparts in Leverkusen.

But this anticipates the story of the politics of coal allocation for the western German chemical industry. It will be useful first to look at Allied policy and its role in the treatment of the I.G. successors from mid-1945 to mid-1948.

ALLIED POLICY TOWARD THE I.G. AT THE SUPERZONAL LEVEL

The three preceding chapters have demonstrated that, during the first months of the occupation, each of the western Allies developed its own style of handling the factories of the I.G. To some extent such policies affected chemical production levels, although the central fact for the chemical industry in all zones in the first occupation years was the shortage of coal. The Allies also attempted to develop superzonal and coordinated policies toward the successors of the chemical firm beginning in late 1945, although success was limited at the four-power level. Not until after the founding of the Bizone (January 1, 1947) did unified superzonal treatment of the firm come about, and then only at the more modest Anglo-American level.

The four occupying powers in Germany agreed to a common policy toward I.G. Farben as of November 30, 1945, when the military governors—Marshal G. Zhukov of the Soviet Union, Field Marshall B. L. Montgomery of Great Britain, General J. T. McNarney of the United States, and General P. Koenig of France—signed into effect Allied Control Council (ACC) Law Number 9. The law provided for the seizure of I.G. assets and their vesting in the Control Council. The specific basis of the seizure was the American argument for elimination of German war potential: the ACC enacted Law No. 9 "in order to insure that Germany will never again threaten her neighbours or the peace of the world, and taking into consideration that I.G. Farbenindustrie knowingly and prominently engaged in building up and maintaining the German war potential."[13]

Law No. 9 also provided for a committee of the four I.G. Farben control officers from the individual zones. The I.G. Farben Control Committee (COIG) met in Berlin and was directly responsible to the Economic Directorate—although it reported at first to CORC (the Coordinating Committee) of the Allied Control Authority, or ACA. COIG's tasks were wide-ranging: it coordinated control efforts with regard to the I.G. holdings in each of the zones; dealt with financial questions relating to those holdings (e.g., taxes, pensions, and so forth); and tried to come up with a coherent and acceptable plan for the dispersal of the giant firm. The deputy controller of the I.G. for each zone had the task of implementing COIG decisions. Meeting for the first time on February 4, 1946, the committee continued to convene at least once a month into March 1948.[14]

Initially, there were some grounds to hope for success. At the first meeting, control officers exchanged information on how Farben was treated in each zone, this in order to provide the basis for more unified control. COIG also began to settle outstanding difficulties extending beyond zonal boundaries in the control of the I.G. Problems of finding operating funds for Ludwigshafen, for instance, were handled by COIG since the central financial offices of the I.G. were not in the French but rather in the U.S. and Soviet zones. COIG agreed to give the French RM 10 million of an RM 17 million request, half from each of the U.S. and Soviet zones.[15] The committee and its staff completed an investigation of all of the assets of the I.G. by October 1946, and COIG was able to come up with an impressive list of the decisions it had taken by February 1947, just under one year from the date of its first meeting. Included were decisions:

(i) To decentralise complete [sic] the management of Plants and enterprises controlled by I.G. Farben.
(ii) To deprive the central I.G. Agencies of all functions of administrative, technical and financial management.
(iii) To liquidate all central selling organisations.
(iv) To prohibit the further carrying out of all I.G. obligations under cartel agreements.
(v) To divide such part [sic] of I.G. which it is desired to keep in existence under the Level of Industries programme into as many separate units, each capable of independent economic operations, as possible.
(vi) To appoint Custodians to hold and administer the separate units until their final ownership is determined.
(vii) To withdraw all Powers of Attorney, and other powers, to sign on behalf of I.G. Farben, from German officials, except where these may be specifically given or reconfirmed by the individual Control Officers.
(viii) To dispose of all I.G. participations not exceeding 25% of the capital of outside undertakings, by sale or otherwise as circumstances permit.
(ix) To dispose of assets, not in use for industrial purposes, by sale or otherwise. . . .[16]

The success of the committee, such as it was, was short-lived. COIG found operating money for the Ludwigshafen plant, but this decision was undermined when the Soviets refused to meet their part of the obligation.[17] Actual control of the I.G. continued to be effective only at the zonal or factory level, because, in fact, little had actually been accomplished at the superzonal level. The first four of the "decisions" listed above were not decisions at all. They were simply admissions, first, of actions already undertaken by zonal military governments with regard to the I.G., and, second, of the *de facto* breakup of the I.G., and especially of its central agencies, *by virtue of the division of Germany* into four relatively isolated zones of occupation. No action had been taken on the final three items listed. And crucial decisions on how many separate units of the I.G. would exist and, more importantly, on the size of a unit necessary for it to be "capable of independent economic operation" were simply not made.

COIG continued to meet into 1948, but with less and less effectiveness. The final blow came in March of that year as the apparent institutionalization of the division of Germany (through creation of the Bizone) and the increasing tensions of the Cold War took their toll on four-power relations. To some extent, though, unified Allied policy had been extremely difficult from the outset given the initial separation of policy formation and its implementation, and the relatively late attempt on the part of the Allies to coordinate policy. Only in early 1947 at the bizonal level did more effective unified action toward the I.G. at the superzonal level begin to emerge.

The beginnings of formal Anglo-American administrative cooperation in Germany lie in the months before the official founding of the Bizone in January 1947. Starting in early fall 1946, central bizonal administration was gradually set up in Berlin. Known as the Bipartite Board, it later set guidelines and made the fundamental decisions in bizonal policy. A set of "panels" stood under the Board; they generated suggestions for policy in specific areas—such as the economy—for the Board. Policy implementation was the responsibility of the Bizonal Economic Control Group (BECG), which stood under the Bipartite Board and was made up of members of British and American military government apparatus controlling various areas of the economy. No single organization or section of the bizonal economic control apparatus concerned itself exclusively with the successors of the I.G. Farben. Rather, responsibility was shared by the chemical industry subsection of the BECG Basic Industries Section, and several other bizonal organizations. The Joint Export–Import Agency (JEIA) controlled chemical industry trade, for instance, and working parties within the BECG Basic Industries Section dealt with overlapping problems affecting a number of industrial sectors. A reorganization of the bizonal administration occurred at the beginning of 1948 with the creation of the Bizonal Control Office (BICO). BICO replaced the Bipartite Board and signified the enlarged competence of the bizonal administration—i.e., extending to more than just economic policy issues. The BECG stood under BICO, and was itself replaced by a new organization, the Commerce and Industry Group, in March 1948.

Bipartite agencies responsible for the chemical industry in the American and British zones were not very effective through mid-1948.[18] Much control over the individual chemical plants, and over the former I.G. Farben factories in particular, remained in the hands of zonal rather than bizonal authorities. On the other hand, bizonal authorities responsible for the industry were generally also top-level zonal officials with competence in the area of the chemical industry. Thus, some coordination of policy and its implementation resulted simply from the providing of a forum where U.S. and British officials could meet.

Two important elements characterized Anglo-American control of the I.G. and the chemical industry in the Bizone. For one thing, jurisdiction among responsible bizonal agencies overlapped considerably. Not until August 1948, when it became clear that four-power control of the I.G. had failed permanently, was the Bipartite I.G. Farben Control Office (BIFCO), set up as an agency with specific competence with regard to the I.G.[19] Even then, responsibility for specific issues such as dismantling, production levels, and foreign trade remained in the hands of other bizonal organizations. In addition, a key aspect of the bizonal organizational program was to hand over considerable power to German authorities. During the initial bizonal period, German trustees for individual I.G. factories exercised far-reaching authority over the plants. In addition, the chemical authorities within the *Verwaltungsamt für Wirtschaft* (Administrative Office for Economics) (VAW)—which later became the *Verwaltung für Wirtschaft* (Economics Administration) (VfW)—were heavily involved in the control and supervision of the chemical industry; they acted as go-betweens to promote the interests of the I.G. successors with the bizonal occupation authorities.[20]

There were close parallels between the bizonal control structure for the chemical industry and the I.G. and the organizational characteristics of the American zone of occupation. Overlapping jurisdictions typified the U.S. control apparatus to a greater extent than that of any of the other Allies. In addition, American zonal authorities tended to turn over large measures of power to Germans, preferring instead simply to supervise their actions. In terms of those characteristics, then, British and American treatment of the chemical industry and of the I.G. successor companies became increasingly unified. What is more, bizonal policy came to resemble more and more already existing policy in the U.S zone.

GERMAN CHEMICAL AUTHORITIES IN THE BIZONE AND THE POLITICS OF COAL

A set of German organizations acted parallel to, and in conjunction with, the Anglo-American organizations responsible for the chemical industry in the Bizone.[21] In late 1946, British and American military government authorities allowed the Germans to set up sections (*Referate*) responsible for industries in the Bizone that would come under the jurisdiction of the VAW. The chemicals section, composed of members of corresponding offices at the British zonal and U.S. zonal and *Länder* levels (see fig. 6), was placed under the direction of Dr. Theurer (section leader), formerly with the *Länderrat* in the U.S. zone, and Dr. Pohland (deputy leader), formerly of the Central Economics Office in the British zone.

From its inception, the chemicals section of VAW had two distinct sets of functions that it inherited from the chemicals section of the Reich Economics Ministry (RWM) and the Reich Office for Chemicals (*Reichstelle Chemie*). The functions were, first of all, "ministerial responsibilities" formerly associated with the RWM, including basic direction of and economic policymaking for the chemical industry; and, second, development of plans, programs, and allocations (*Bewirtschaftung*) for the industry, a task formerly performed by the *Reichstelle Chemie*. By the end of 1947, the VAW

DIVISION LEADER: Dr. Theurer

Deputy Division Leader: Dr. Pohland

Section/Leader	Areas of competence
I. **Central Section** (Dr. Pohland)	a. coordination b. **dismantling questions** c. planning d. statistics e. documentation
II. **Trade** (Bünger)	a. foreign trade b. interzonal trade c. price formation
III. **Chemical Engineering** (Dr. Hauck)	
IV. **Inorganic Products I** (Dr. Uebler)	
V. **Inorganic Products II** (Dr. Hoffman)	
VI. **Organic Products I** (Dr. Blankenfeld)	
VII. **Organic Products II** (Dr. Petrasch)	
VIII. **Consumer Chemicals I** (Dr. Janssen)	
IX. **Consumer Chemicals II** (Dr. Lederer)	

Figure 6: Organizational Plan of the Chemical Division in the Administrative Office for the Economy, Minden (as of September 1, 1947).

Source: Adapted from Friedrich Stratmann, "Chemie unter Zwang?" (Dissertation, University of Göttingen, 1984), Anhang ABB. 3.

chemicals section formally divided these responsibilities once again, retaining the ministerial functions and setting up a subsidiary allocations office (*Bewirtschaftungsstelle*). Personnel responsible for administration of the two areas were often the same, however, and thus the functions remained united for all practical purposes. Responsibilities for detailed planning and allocations of course gradually disappeared after the liberalization of the West German economy following the currency reform.[22]

There was yet another set of organizations responsible for the industry in the Bizone. The Provincial Committee for Chemistry (*Länderausschuss Chemie*, or LA) was composed of representatives of the VAW chemicals section as well as their counterparts at the *Länder* level in the British and American zones. A parallel organization, known as the Specialized Committee for Chemistry (*Fachausschuss Chemie*, or FA), was made up of VAW representatives, leaders of bizonal chemical industrial organizations (*Wirtschaftsverbände*), and personnel from the newly formed chemical-industry trade unions. Beginning in August 1947 and continuing into the federal period, the two committees met together as the *Länderfachausschuss Chemie* (LFA).[23] Composed of twenty-five to thirty persons representing government, industry, and labor in discussions on issues related to the chemical industry, the LFA was

> the decisive steering and consensus-building center of the chemical economy above the factory level. If, too, officially no "decisions" were made or were supposed to be made in the form of votes and it was stressed again and again that the body should be "advisory," the decisions reported in the minutes may be seen as consensus decisions [*Konzensbeschlüsse*] in all important questions.[24]

The LFA Chemie thus served as a forum in which to discuss and unify policy toward the chemical industry (and toward the Allies in areas affecting it). It became a bridge not only between central and regional authorities responsible for the industry in the Bizone, but also between government officials, industrialists, and representatives of labor.

The VAW chemicals section and the LFA Chemie dealt with the individual firms of the former I.G. and their representatives in the process of discussing and acting on a wide variety of issues.

Still, until after the currency reform, there was no German governmental body in the Bizone responsible exclusively for issues connected with the I.G. Individual Germans, of course, served as "Trustees" for the Allies in the administration of individual I.G. plants, and some Germans—such as Paul Moldenhauer[25]—served as consultants to the Allies in developing policy toward Farben even in the initial years of the occupation. But the I.G. did not lie within the official competence of the VAW and associated bodies. Only in August 1948 did a German body responsible for suggestions with regard to the I.G. come into existence. The I.G. Farben Dispersal Panel, or FARDIP, as the new body was called, was nongovernmental in its composition, although VfW officials made recommendations on its personnel.[26] It was created as a counterpart to the Anglo-American BIFCO, and as a response to the apparent futility of four-power policy toward the I.G.

In trying to expand—or even to maintain—production of chemicals through mid-1948, representatives of the German chemical industry confronted a large number of serious problems. Lack of adequate means of transport, caused by damaged bridges and railroads and blocked rivers, meant that plants often received raw materials only with great difficulty. Finished products often could not be shipped to customers for the same reason. For some of the former I.G. companies, and especially for the Farbwerke Hoechst, the breakup of the chemical combine had serious ramifications in terms of supply. Early in 1948, for instance, Hoechst officials complained that they could not get enough acetaldehyde—an important intermediary in the production of several organic chemicals—from their former supplier within the I.G., Knapsack. The reason given was that "out of pricing considerations, Knapsack is not interested in supplying Hoechst directly with acetaldehyde."[27]

In terms of the number of firms in the chemical industry affected, short supplies of various vital raw materials were much more serious than the problems associated with the breakup of Farben. In Hesse in February 1948, for instance, a shortage of a seemingly minor product, paper sacks, prevented many chemical firms from filling fertilizer orders.[28] At about the same time, limited availability of nonferrous metals, crucial as catalysts for many chemical processes, took on "catastrophic forms" in the Hessian chemical industry. The shortage was noted in a report by the LWA Hessen, which went on to warn:

> If, therefore, these metals are available only in homeopathic quantities [*homöopathischen Mengen*] or in very many cases not at all, the point will be reached in the shortest time at which all will simply come to a standstill.[29]

In attempting to solve these and other severe problems, German governmental agencies performed an active and often vital role. The discussion of the problem of acetaldehyde supply to the Hoechst factory took place in the LA Chemie and resulted in a meeting between representatives of Hoechst and Knapsack and Herr Duerholt of the LWA Hessen and Dr. Heubaum of the NRW Economics Ministry to work out a solution to the problem.[30] The agencies developed plans to try best to "administer shortage."[31] But the most important brief for German governmental authorities during the period 1945 to 1948 was to solve the problem of coal supply to chemical plants. By examining this area, the interactions of all of the actors interested in chemical production in the Bizone are thrown into sharp relief.

Coal supply played a much bigger role in the mid-1940s than it does today in the chemical industry. Used as a raw material in its various forms, coal was also the most important source of fuel for industry. Coal tar served as the basis for a vast palette of synthetic dyes and desperately needed medicines. It drove the machinery and fired the reactors used in chemical manufacture. In the especially harsh winters of 1946–1947 and 1947–1948, it helped maintain (if available in sufficient quantites) sufficiently high temperatures in chemical plants to prevent frost damage. In its absence, production machinery suffered.

Precisely in these initial years of the occupation, however, coal production in the Ruhr area was most severely affected by wartime damage and occupation policy. Destruction to the transportation system meant that coal often lay unused at mine sites, while the insubstantial diet of miners discouraged productivity. Wartime losses meant that labor was in short supply and advanced in age. Housing for miners was inadequate. Initial occupation policies exacerbated the problems: restrictive regulations curbed productivity; dismantling of coal mining equipment impeded production.[32] The result was poor performance in the western German coal industry into the early 1950s. Even in 1950, when West German industry in general was producing at 140 percent of its 1936 pace, the coal industry had just reached its 1936 rate.[33] Thus, despite changes in Allied policy—toward using incentives of extra food and clothing to encourage miners in the Ruhr fields to be more productive, and decreasing exports of German coal—coal supply remained one of the most intractable problems facing the postwar German economy.

German and Allied military government agencies responsible for control and direction of the chemical industry tried valiantly to aid it in developing ways to overcome the problem. The obvious answer was simply more coal production, since it would of necessity mean more supplies to the chemical industry. Incentives to miners tended to be successful only in the short term, however, and transportation outages and shortages of pit props and other items continued to thwart both miners and authorities. When increases in production proved temporary, various agencies drew up plans to distribute available coal resources better and more effectively.

Industry and its representatives, for instance, worked with the Chemicals Section of the VAW in the summer of 1947 to try to stockpile coal. A report from the section for that August noted such an attempt by chemical firms in the combined area, despite serious supply problems, because of their experiences in the preceding winter. It went on to point out that progress toward this end, though noticeable, was insufficient, and thus discussions were underway within industry to develop plans to shut down chemical plants for several weeks in the autumn. Coal would consequently be available for heat and production in the hardest winter months, the main objective being the prevention of frost damages.[34]

The plan suffered from two obvious drawbacks. It was, first of all, difficult for plants to stockpile coal when sufficient supplies even for minimal levels of production were not forthcoming. Furthermore, given interdependence in the chemical industry, and the dependence of numerous other industries on it, disruption of production "for several weeks in the autumn" would have obvious ill effects on overall bizonal production levels.

Less obvious, but just as important, was an objection that came up at an LFA Chemie meeting in August 1947. On the eighth of the month, committee members discussed the relative merits of stockpiling. It was, they thought, basically a good idea. But even if the plan succeeded, it appeared likely that the industry's coal reserves would be seized to supply hospitals and the food-processing industry in the event of a hard winter. Consequently, they agreed to press for approval from the upper ranks of the VAW and from the military government to establish a system of "coal protection letters" (*Kohlenschutzbriefe*) for firms in the industry. The idea was to provide companies with an incentive for stockpiling coal by giving them certificates that would guarantee against seizure of reserves.[35]

The plan went no further than discussion and resolution in the LFA Chemie. In any case, this and other plans and suggestions related to stockpiling and increased production ran into a fundamental reality of the occupation: there simply was not enough coal to go around. What is more, the structural reasons for underproduction were going to take a long time to overcome. Two other types of plans for overcoming the problem of coal supply to the chemical industry through the middle of 1948 attempted to deal with this fundamental reality. While only one was employed, they shared a common thrust: both were directed toward maintaining or improving coal allocation to large firms in the industry rather than to smaller ones.

The first promoted the idea of "full utilization of reserves in short supply in the combined zones." Advanced by the Anglo-American Bizonal Economic Control Group and sent to the Chairman of the VAW on May 20, 1947, the plan sought to prevent waste of resources. It called for preference in allocations of raw materials to more efficient firms while at the same time providing for equity in allocation of finished products:

> We cannot afford the luxury of maintaining any kind of predetermined production relationships between the two halves of the Combined Zone or between the *Länder*. The allocation of raw materials and semi-finished products must be built upon the principle of the greatest possible utilization of production capacity. The allocation of end products must be based on the principle of an equal satisfaction of demand in all parts of the Combined Zone.[36]

Efficiency "in the best factories," the memorandum continued, was not necessarily the province of large firms only, "although this seems to be the case very frequently." In any case, firm size as well as location of the plant in the British or U.S. zone or in a certain *Land* was to play no part in allocations. Rather, in language reminiscent of the period from which Germany had just emerged, the memo maintained that "every plant must demonstrate its ability to survive through its economic and social achievements."[37] The memo did not call for immediate implementation of the plan, but rather for suggestions on how to do so from German authorities.

The Allied request met with little sympathy or interest from the German side. The chemical industry's trade association in the British zone, for instance, contended based on its own studies "that, in the area of the chemical industry, concentration of manufacturing through coercive measures could not be implemented," and it lobbied against the Allied effort on this basis.[38] The plan was never implemented. But it showed that, ironically, the same Allied authorities who were interested in preventing the concentration of economic power in Germany were prepared to forego that goal when confronted with a problem as pressing as coal supply.

In addition, the Allied plan may have played a role in the development of an alternative program by German authorities for the solution of the coal-supply problem in the chemical industry. This plan, which *was* implemented, also favored larger firms, albeit in a less obtrusive way. Before discussing this plan in detail, however, it will be necessary to look at the process by which bizonal coal allocations were made through mid-1948. Throughout the bizonal period, the VAW (later the VfW) was responsible for global allocations: on the basis of information provided by the *Länder* and by the agency's industry subsections, the VAW allocated coal and other resources to civilian, institutional, and industrial consumers. Military government officials then could exercise the right of veto over the allocations, although in general they made only minor changes.

Allocations were classified in two ways. First of all, there were central allocations—to the Reichsbahn and to hospitals, for instance—which were determined by the VAW and thereafter unalterable. Most industries, in contrast, were allocated "globally" by province. Certain amounts of coal were set aside for each industry in each *Land*, depending upon the industry's significance and upon its priority in the Bizone as a whole. *Länder* representatives themselves divvied out coal to particular firms, since the VAW believed that they were most familiar with local conditions. What is more, they had the power to alter VAW allocations to a particular industry by 10 percent.

That the chemical industry suffered under this system was clear to German authorities at the bizonal level. A report delivered in October 1947 noted that the industry was faced not only with consistent underallocation of coal, but also with consistent underdelivery of its allocation. In fact, chemical producers as a whole received more than their stated allocation of coal in only one month—July—between April and August 1947, when 112 percent of the promised total arrived. In other months, deliveries ranged between 80 and 86 percent of allocations. Underdeliveries, the report went on, were never made up since the slate was wiped clean at the beginning of each month. The reason for underdelivery was clear: it rested solely on the discretionary power of the *Länder* to alter allocations to most industries by 10 percent. That power was almost always used, and, as the figures for delivery made clear, at times even more was cut.[39]

Authorities responsible for dealing with the chemical industry in the Bizone had begun to act upon this problem even before the presentation of the report. The impetus behind the effort appears to have been the problems of coal supply faced by the former I.G. plant at Leverkusen. The plant shut down completely from December 1946 through February 1947. By March, production began again slowly as coal stocks increased. Almost immediatelty thereafter, Leverkusen's representatives began to agitate for larger allocations. Dr. Oskar Loehr of I.G. Leverkusen, in a letter of Dr. Keiser of the VAW Planning and Statistics Section on April 2, 1947, complained that "the provision of the Leverkusen factory with coal has got so much worse since the beginning of this year that secure production is no longer possible." Coal deliveries for a given day, Loehr continued, were less than the plant's requirements for that day. If deliveries were delayed (as they often were), the result was often production cutbacks, loss of valuable raw materials (since slowdowns, shutdowns, or variation in temperature often rendered them useless), and, in winter, frost damage. He asked for extra allocations of coal to allow the plant to stockpile. In this way, fluctuations in production levels could be avoided. As justification for his request, Loehr pointed to the overwhelming importance of the plant to the industrial economy. He noted: "the Leverkusen factory represents with its approximately 15,000 employees the largest industrial plant at the present time in the British and American zones." The plant supplied around 1000 chemical products, most of which served as intermediate materials for other industries, and it contributed heavily to the bizonal export program.[40]

Loehr and his colleagues at Leverkusen requested an allocation of 20,000 hard coal units (HCU) per month, approximately what the plant had received during the war, and much more than its usual monthly allocation in 1947 of 9,000 to 12,000 HCU per month. VAW personnel were sympathetic but questioned the need for coal provisions similar to those during the war at a time when production stood at about 40 percent of wartime levels.[41]

Leverkusen's coal expert, Dr. Warnecke, justified the request in a letter to the VAW at the beginning of May. Assuming that the factory would produce for the remainder of the year at 62 percent of its 1939 rate, its coal needs would be approximately 15,000 tons per month. Warnecke contended, however, that changed production relationships since 1939 meant that in fact more coal was needed in 1947 to produce at a rate similar to the earlier period. He itemized the additional requirements and their justifications as follows:

additional requirement	justification
1200 t.	inefficient firing in existing equipment
900	poor coal quality
2000	problems with poor insulation, ventilation
500	building without roofs and windows
1100	inadequate public supplies of electricity and coal gas force self-generation
2000	to make up for past production losses.

The sum of projected needs based on 1939 and additional requirements was 22,700 tons, and this, Warnecke held, justified fully the firm's request for 20,300 tons.[42]

Sympathy with the predicament of the factory and desire to improve Leverkusen's position on the part of German bizonal authorities did not at first translate into increased allocations. In fact, in early June 1947, factory management at the former I.G. plant decided to shut down the plant once again. Once again it was owing to coal shortages. Coming as it did such a short time after the previous one, this production stop galvanized the VAW into action. The first member to respond was Dr. von Maltzan, head of the important Main Division for Export and Interzonal Trade. Von Maltzan wrote to Leverkusen on June 16 to express his regret that the pessimistic predictions of company representatives had indeed come to pass. He continued, stressing the importance of the plant to efforts to increase bizonal exports:

> I regret this all the more as I am conducting the first especially promising negotiations with the leadership of your firm as I begin my efforts to revive German export, and from that had been able to gather the expectation that one could reckon in the chemical sector with a slow but sure rise in exports from current production.

In closing, von Maltzan offered to do all in his power to aid the firm in returning to production. He wanted, however, a clarification as to whether the coal supply problems stemmed from low allocations or from low deliveries. Believing the latter to be the case, von Maltzan asked if, perhaps, the firm could collect its own coal since it lay in nearby coalfields.[43]

Major action to solve the coal supply problem at Leverkusen once and for all came later in June, however, at a meeting between top officials of affected departments in the VAW and top representatives of the I.G. Leverkusen plant. The former included Dr. Keiser of the Planning Section, who chaired the meeting, and Drs. Pohland and Hauck of the Chemicals Section. Leverkusen's representatives included the plant manager, Dr. Ulrich Haberland, and Mr. Tietz. From the beginning of the meeting, Dr. Keiser let it be known that he was willing to use all of the resources at the VAW's disposal to get the plant back into production. Saying that "the shutdown of Leverkusen is insupportable for the entire economy, and the factory because of its super-regional significance could no longer be left to the exclusive control of the Land of North Rhine Westphalia," Keiser simply asked firm representatives what in their opinion would be necessary to bring the plant back into production.

Haberland, speaking in response, delineated the coal supply

situation of the firm. He had closed the plant down on June 4 because supplies of coal had run down to just 80 tons. He and his fellow factory managers had decided, furthermore, not to resume production until 6000 tons of coal were on hand. Current supplies amounted to about 3000 tons. At this point, Dr. Hauck of the chemicals section of the VAW took up the tale. His section had carefully scrutinized Leverkusen's requests for around 21,000 tons of coal per month and had found the magnitude of the proposed allocations a bit high. He agreed, however, that the crux of the company's argument was correct: current diseconomies meant that allocations would have to be higher than would be the case under normal conditions. He thus suggested that a more reasonable allocation figure would be around 18,000 tons per month.

Keiser then took the floor again to propose his plan of action. Leverkusen should receive what it was owed from previously underdelivered allocations from North Rhine Westphalia. What is more, he would put 4000 tons of coal export reserves at the plant's disposal and suggested that, beginning in August, the factory should receive about 15,000 tons HCU per month. Finally, he promised that "from August, the 10 percent deduction for the reserves of the *Länder* should be done away with in the case of the chemical industry." Keiser then asked Haberland if this would be an acceptable basis on which to resume production at the Leverkusen plant immediately. Haberland agreed.[44]

Since the emergency at the Leverkusen plant was dealt with to everyone's satisfaction, representatives from the VAW and the firm turned to exploring more long-term solutions to the chemical industry's supply problems: "It was agreed further that the chemicals section should bring about the central allocation of coal contingents for certain factories of super-regional significance in the *Länderfachausschuss*...," and the chemicals section was directed to come up with allocations for August–September 1947 on this basis. The plants affected were: A.G. für Stickstoffdünger (Köln–Knapsack), Chemische Werke Hüls (Marl), Deutsche Solvaywerke (Rheinberg), Dynamit A.G. (Troisdorf), I.G. Leverkusen, Matthes & Weber (Duisburg), Dr. Alexander Wacker (München), I.G. Hoechst, and Kali-Chemie (Heilbron).[45] Almost all were former I.G. plants or subsidiaries.

The meeting on Leverkusen's coal situation thus had considerable significance. Top representatives of the VAW indicated that they were willing to do whatever would be necessary to resume production at a major chemical plant. They also resolved to try to wrest control for allocations to the industry, or at least to its major plants, from provincial regimes. Finally, the VAW representatives, by extending the solution applied to Leverkusen to all major chemical industry plants in the Bizone, underscored their commitment to plants that employed large numbers of people, produced on a large scale, and/or were important to export, at the expense of smaller plants. Once again, confrontation with the realities of the occupation—the need to maximize German production to meet internal demand *and* to export—overcame plans for severe deconcentration of industry.

By December 1947, the VAW had attained its goal of central allocation of coal to the chemical industry and to certain key firms, although not without considerable frustration and opposition. Despite centralized allocations, it was often not possible to supply key plants with their coal needs. At Leverkusen, for instance, autumn 1947 was a period of extreme uncertainty over the coal situation. At the beginning of October, factory managers believed that they would shortly be forced to shut down once again. By October 21, circumstances had changed because the firm assembled seven days' reserve stock of coal. Company representatives could look by mid-November with confidence toward the rest of the year and believed that they would certainly be able to keep the plant in production through December. By December 9, though, Leverkusen's coal supplies were once again insufficient for a single day of operation.[46] Government officials from the *Länder* of the U.S. zone, what is more, opposed fervently the VAW's centralization of control over coal supplies. At a meeting of the coal subcommittee of the *Länderrat* of the U.S zone on December 5, 1947, officials complained about the centralization of coal allocation and planning:

> VfW Frankfurt has now announced on the alleged instruction of the BECG that, on the subject of the planning for January and February, in addition to the existing consumer groups planned centrally in Minden (electricity works, iron and steel, fertilizer industry, non-ferrous metals, fireproof stones [feuerfeste Steine]), the following groups should be centrally controlled as well: petroleum industry; stones and earths; cellulose, paper, and pasteboard; basic chemical materials in the chemicals group; rubber and plastic.[47]

They claimed in addition that the practice of central allocation eroded the competence of the provinical coal authorities (*Kohlenreferate*) and meant a falling out of touch with conditions at individual factories. They recommended instead that central planning be limited to transportation, electricity works, the iron-and-steel-producing industry, nonferrous metal works, and fertilizer. All other groups should feature centralized suggestions but decentralized control.[48]

By spring 1948, weather conditions improved and the VfW's control of coal allocation for the chemical industry at the central level became consolidated. The *Arbeitsstab-Kohle* (Coal Working Staff) in the U.S. zone dissolved and turned over its responsibilities to the VfW by April,[49] thus eliminating a major rival to centralized allocation authorities. Coal supplies also improved. The chemical industry in the Bizone had been allocated a total of 492,720 tons of coal in the fourth quarter of 1947, which rose slightly to 523,590 tons in the first quarter of 1948. Allocations increased substantially in the second quarter of 1948, though, to 610,000 tons.[50] Since, moreover, they were now controlled centrally, deliveries corresponded more closely to allocations.

Constant crisis within the chemical industry characterized the period from the beginning of the Bizone in January 1947 to the currency reform. Because of the nature of the crisis, government officials found it necessary to intervene. Military government officials were of some importance in this, if chiefly by virtue of the fact that they allowed German officials to pursue their own policies. German administrators were much more active, meeting with personnel from firms, devising emergency solutions to an array of problems, and attempting to come up with long-range solutions to difficulties facing all industrial groups. Both Allied and German officials, however, tended to prefer larger, more efficient firms. Bizonal coal shortage in 1947 and 1948 were primarily responsible. After the currency reform, continuing difficulties in coal supply and other pressing considerations pushed policymakers in the same direction.

6
CONSOLIDATION OF RECOVERY MID-1948–1950

In early 1948, observers of the German chemical industry agreed that its prospects were bleak. British officials responsible for its control in North Rhine Westphalia pointed out in February 1948 "that a considerable decline in the morale of the factory management and workers had occurred during the last year."[1] Outsiders echoed the assessment and held that it boded ill for the future. Norman A. Shepard, the technical coordinator of American Cyanamide Co., concluded a survey of the state of German chemical manufacturing by claiming that "all in all, the German industry is failing to contribute to European recovery."[2] Later in 1948, the director of research at the Hercules Powder Co., Dr. Emil Ott, told the U.S. Synthetic Organic Chemicals Manufacturers' Association that "Germany can never regain the dominant position it once held as a chemical producer in world markets." What is more, Ott continued, "German industrialists have conceded [this]."[3]

Each of these sober analyses of the prospects of the German chemical industry contained a hint of optimism, however. British chemical control officers in North Rhine Westphalia also reported that "in spite of conditions . . . the managements, in particular of larger firms, are still keen and trying hard to maintain and increase their production."[4] Shepard commented on the interest of potential customers in the industry's products; the fact that a "comparatively large amount of chemical capacity . . . had escaped bombing"[5] meant that it might be possible in the future to satisfy them. Ott concluded that the German chemical industry would "not regain its former preeminence in the chemical field, [but] . . . will again become a significant factor" in the world market.[6]

The ambiguity in the analyses was well-founded. The industry had in fact already made significant strides toward recovery by early 1948, although coal shortage, tight supplies of important raw materials, and bureaucratic constraints impeded production and export. The events of mid-1948 were to consolidate the recovery. The first influx of Marshall Plan aid, western German currency reform, and gradual deregulation of the economy had generally positive effects on all industry branches. For the former I.G. factories in particular, the end of the Nuremberg trial of former Farben executives, the establishment of Allied and German committees to oversee the final breakup of the concern, and increasing French participation in policymaking allowed gradual normalization of management and production relationships.

Taken together, the factors marked a turning point in the history of the postwar western German chemical industry. Its fortunes rebounded quickly and significantly. By 1950, production had once again reached—and in some areas surpassed—the levels of 1936. German industrialists and representatives of the emerging state apparatus in western Germany had more and more to say about how the industry was to be organized. Finally, there was an increasing tendency toward three-power unity in western Allied policy toward the German chemical industry. Problems continued, though, in several crucial areas, including coal allocation, investment, and the breakup of the I.G. It would not be until 1951–1952 that resurgence of the industry could begin.

THE WESTERN GERMAN CHEMICAL INDUSTRY AND THE "WIRTSCHAFTSWUNDER"

Something important happened to the western German economy between June and July 1948 (see table 8). Production of bizonal industry as a whole increased slowly but steadily throughout 1947, but during the first half of 1948 production stagnated at about one-half its 1936 rate. In the single month from June to July 1948, however, output jumped about 20 percent. Virtually every industrial sector featured significant production increases. The production jump inaugurated a period of sustained and spectacular growth in the West German economy that continued with some minor interruptions into the 1960s, and thus has come to be seen as the beginning of the West German *Wirtschaftswunder*, or economic miracle.

The causes (and thus ultimately the significance) of the German

Table 8. Development of Production for Selected Production Areas in the Bizone, 1946–1948 (1936 = 100, not seasonally adjusted)

	All Industry	Coal	Iron & Steel Products	Machine Tools & Optical Industry	Chemicals Industry	Textiles & Clothing
Weight	100	8.6	8.7	9.9	8.6	11.0
1946	34	51	20	35	39	20
1947	40	65	22	34	40	30
1948	60	79	38	52	61	52
1947						
Jan.	30	60	17	23	30	23
Feb.	29	59	16	23	28	21
Mar.	34	66	21	25	37	28
Apr.	39	56	22	33	45	31
May	41	58	22	39	46	30
Jun.	41	61	21	40	42	30
Jul.	42	67	23	35	46	32
Aug.	42	68	24	32	44	31
Sep.	43	69	24	36	39	32
Oct.	45	73	26	41	41	34
Nov.	45	72	26	40	42	33
Dec.	44	72	26	40	43	30
1948						
Jan.	47	73	27	40	46	36
Feb.	47	70	26	44	48	39
Mar.	51	78	29	45	53	41
Apr.	53	76	31	46	56	44
May	47	66	27	41	52	38
Jun.	51	80	32	43	53	44
Jul.	61	83	40	45	64	54
Aug.	65	81	44	53	68	61
Sep.	70	82	47	60	70	63
Oct.	74	86	51	65	72	67
Nov.	76	84	51	73	75	67
Dec.	79	89	55	73	76	69

Source: Adapted from Werner Abelshauser, *Wirtschaft in Westdeutschland 1945–1948* (Stuttgart: Deutsche Verlags-Anstalt, 1975), p. 43.

economic miracle are less readily apparent. Accepted wisdom stresses three interrelated factors as the driving forces behind the *Wirtschaftswunder*.[7] The proximate cause was the enactment on June 20, 1948, of a currency reform in all three western zones of occupation. With one stroke, the reform eliminated the discredited German currency, the Reichsmark, and substituted for it a new one, the deutsche mark. Most old currency had to be converted at the average rate of RM 100 to DM 6.5 (although a small amount was exchanged at 1:1). The result was electrifying:

> What seemed almost incredible happened. Literally from one day to the next fresh vegetables appeared in the windows of food stores empty for years; shoes, clothing, and underwear, unobtainable for money the Saturday before, could once more be bought. It now made sense to supply the markets with hidden goods and subsequently to produce goods for those markets.[8]

West Germans responded positively to the reform since it replaced the old currency—which was thoroughly discredited by wartime and postwar inflation—with one more realistically pegged to domestic and international currency markets. Of course, few Germans as yet had anything to do with international markets, but the psychological effect of the long-awaited and generally applauded action was substantial.

According to the traditional interpretation, the currency reform would have been largely ineffective had it not been for two accompanying factors. For one thing, monetary reform had to be coupled with moves to liberalize and deregulate the economy. Ludwig Erhard, the head of the economics section of the *Verwaltung für Wirtschaft* and later West German economics minister, began this process in late summer 1948 by pressing for decontrol of all but the most essential foodstuffs and raw materials.[9] The final impetus toward the *Wirtschaftswunder* was the European Recovery Program (ERP), or Marshall Plan. In April 1948, the western zones of occupation in Germany received an invitation to participate in the ERP; desperately needed dollars began to finance imports of foodstuffs and raw materials at about the time of the economic reforms of the summer of 1948. The program thus provided vital inputs into the western German economy and helped ease its balance of payments difficulties. In the traditional view, the ERP was the foreign trade component of the general economic stabilization and return to productivity brought on by currency reform and decontrol of the economy.[10]

More recently, students of the German economic miracle have revised this view.[11] Proponents of the revisionist viewpoint contend that, at the latest, recovery in all western zones of occupation began in earnest in 1947, and indeed largely because the German industrial plant was far more intact after the war than one might have expected. Furthermore, bizonal production statistics tended to underestimate recovery through June 1948 (since much of the economy functioned "underground") and to overestimate it thereafter (since goods produced prior to the reforms surfaced after they took place). Of course, problems continued in transportation and energy supply, and they held the economy in check in early 1948. Thus, the economic reforms and the Marshall Plan helped to ease the problems and to sustain the already incipient recovery. The three traditional factors were important, but they did not *cause* the recovery. In general, according to the revisionists, the significance of the three has been overemphasized.

Both the traditional and the revisionist interpretations accept that the Marshall Plan, the currency reform, and the beginnings of a market economy were crucial in determining the pace and direction of the German economic miracle. They differ on the issue of whether these factors caused the miracle or merely accelerated ongoing recovery. How is one to decide between them? Since the most recent work on the subject, by reaffirming the traditional interpretation, indicates that the debate may become circular,[12] a new approach is in order. Rather than concentrating on macroeconomic developments (as do proponents of both interpretations), one may look at a single branch of the economy. The development of the chemical industry during 1948 suggests some new ways of understanding the nature and causes of the economic miracle.

The trend of industrial production for the chemical industry in the initial postwar period was similar to that for all industry (see table 8). Production rates rose slowly but steadily until late 1947, when they stalled for six months at around half the 1936 rate. For the chemical industry, as for industry as a whole, the production index jumped about 20 *percent* from June to July 1948, and continued to climb steadily thereafter. By March 1949, it reached

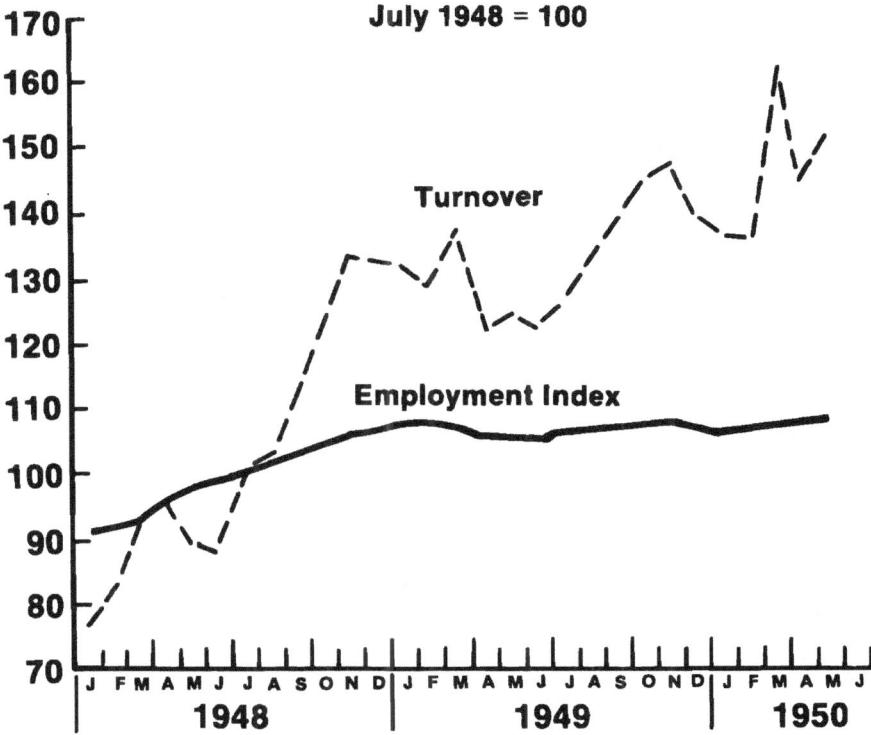

Figure 7: Turnover and Employment in the Chemical Industry of the Bizone.

Source: A.G. Chemische Industrie, "Graphische Darstellungen aus der fachpolitischen Tagung der A.G. Chemische Industrie am 6. Juli 1950" (Archiv des Verbandes der Chemischen Industrie, Frankfurt).

nearly an index level of 90 as compared with the 1936 rate of production.

Figure 7 gives a different, longer-term perspective on post–currency reform trends in the chemical industry in the Bizone. German chemical companies clearly sold considerably more goods at the end of 1948 than they did at the beginning of the year. Corporate turnover rose from January to April before leveling off through June. More substantial growth rates coincided with the currency reform. Turnover at the end of 1948 was more than 40-percent higher than it had been in June. The trend during the first months of 1949 differed from that for production, however. In fact, the first six months of the year saw at first a leveling-off and then a downturn in sales turnover for chemical firms. Growth continued, albeit with interruptions and at a slower rate, in the second half of 1949 and into 1950. Mid-1948 represented no more than a temporary positive change in growth patterns for the chemical industry. What is more, significant problems continued to plague it.

Employment trends for the bizonal chemical industry from January 1948 to mid-1950 demonstrated that production growth did not necessarily mean new jobs, at least during the early years of the *Wirtschaftswunder*. The number of employees rose with economic recovery (although at a far lower rate) from an index of just over 90 (with July 1948 = 100) to about 108 at the end of the year. Employment levels fluctuated somewhat thereafter, but stayed at an index level of between 105 and 108 through mid-1950.

Furthermore, the production index for the bizonal chemical industry itself concealed important subtrends (see table 9). A look at the construction of the production index is necessary to explain this. Occupation officials chose 1936 production levels as the basis of the index since it was considered a relatively "normal" year for the economy: Germany had overcome the Depression for the most part, but armament expenditures had not yet begun to distort production levels in some sectors. They then estimated production of various products in the bizonal area (which of course did not exist in 1936). Thereafter, they assigned weights to individual industrial sectors, and to individual product lines, based on an evaluation of their relative economic importance. Volume of chemical production, for instance, received a weighting of 8.58 in the industry index. The monthly average production of coal-tar dyes in metric tons received weighting of 1.71 of the total index and constituted almost 20 percent of the chemical index.[13] The products listed in table 9 together constituted 76 percent of the chemical index and thus accounted for most of its fluctuations: in other words, variation in the production index for these products constituted more than three-fourths of the average monthly variation in the production index for the bizonal chemical industry as a whole.

The important jump in the index of industrial production in the Bizone occurred from June to July 1948, when production volume increased by about 20 percent. Chemical production on the whole

Table 9. Production of Important Chemicals in the Bizone, 1936, 1946– March 1949 (1936 = 100)

	Nitrogen (fertilizer)	Sulphuric Acid	Calcium Carbide	Soap/Washing Powder	Paint/Varnish/ Laquers	Coal Tar Dyes
1936 m.a.[a]	100	100	100	100	100	100[b]
1946 m.a.	50	34	49	36	n.a.	n.a.
1947 m.a.	63	63	45	33	n.a.	10
1948 m.a.	95	76	70	40	73	32
1948 Jan.	85	63	63	32	45	19
Feb.	80	63	61	30	51	26
Mar.	89	70	70	29	52	29
Apr.	88	71	71	37	58	29
May	89	73	71	30	54	23
Jun.	90	70	77	32	52	19
Jul.	99	76	87	41	65	32
Aug.	107	76	95	42	88	35
Sep.	101	77	82	50	85	39
Oct.	99	86	60	50	112	42
Nov.	101	88	48	57	109	39
Dec.	107	93	52	48	101	45
1949 Jan.	113	104	62	55	97	45
Feb.	111	95	55	50	96	42
Mar.	128	109	87	49	105	48

[a] m.a. = monthly average.
[b] Note that production of coal tar dyes in 1936 was just 3100 metric tons per month on the average.

Source: Calculated from OMGUS, Report of the Military Governor, Statistical Annex Number 45 (March 1949), p. 121.

Table 10. Increase in Production Index for Key Chemical Products, June–July 1948

Chemical Product	Percent Increase
Nitrogen fertilizers	+10
Sulfuric acid	+ 9
Calcium carbide	+13
CHEMICAL PRODUCTION INDEX	+21
Soaps/washing powder	+28
Paints/lacquers/varnishes	+25
Coal tar dyes	+63

Source: See table 9.

went up 21 percent. Although production jumped across the board from June to July 1948, the extent of the increase depended upon product: output of nitrogen fertilizer, sulfuric acid, and calcium carbide lagged, while production of the other materials exceeded the average rate (see table 10). A number of factors accounted for differences between the two groups. Products with increases below the industry average tended to have monthly absolute indices above average, while the reverse was true for those products with increases above the industry average. What is more, products with lower increases tended to be intermediate products, the market for which was other producers. Some nitrogen production was, of course, included directly in finished fertilizers and thus violated this rule. Products with above-average increases tended to be consumer products. Again there was an exception: most dyes were probably sold to the textile industry.

Analysis of these production increases allows important conclusions on the effects of currency reform and liberalization of the economy on the bizonal chemical industry. Overall, the measures had a positive effect, but they also tended to equalize production levels in various product groups in relation to 1936: those products with high absolute indices increased less significantly than did those with low absolute indices. More importantly, it is clear from this limited sample that the initial postreform period tended to favor consumer goods. Low production rates across the board for these three items before July 1948 indicated, furthermore, that there was perhaps less demand for these products before the reforms. More likely, chemical firms chose not to produce such items before the reforms since revenue would have been in nearly worthless currency. Finally, trends in the chemical industry indicate that intermediate goods had already reached relatively high production levels by the time of the reforms and increased less significantly as a result of them. In the case of nitrogen fertilizers, relatively high production levels were the result of Allied occupation policy. In the case of other products, early "recovery" may well have been due to the fact that they could be traded to other producers for much-needed goods[14] or advanced to other producers on the promise of future payment in more stable currency. Such a conclusion would tend to support the thesis of revisionists that recovery preceded the reforms.

The reforms of mid-1948 were thus important in explaining economic recovery, but they apparently had a greater effect in some areas (i.e., consumer goods) than others (i.e., producer goods).[15] Yet another conclusion emerges from this analysis: there are limits to what quantitative evidence can tell us about the origins of the *Wirtschaftswunder*. Again, concentration on a single branch allows an advantage over macroeconomic analysis, since qualitative evidence supplements statistical material. The effect of previous knowledge of the currency reform and deregulation on postreform growth, and the experiences of individual firms in the immediate period following the reform, are important areas to consider.

From the end of the war, all observers were certain that a revaluation of the German currency would have to take place. Wartime expenditures that far exceeded government income and growing shortages of consumer goods combined to cause runaway inflation of the reichsmark. The totalitarian Nazi regime—and in the short run the Allied military governments—could avoid the worst consequences of the inflation through strict rationing of goods. By 1948, however, the reichsmark retained little value beyond its use in buying rationed goods. For further transactions, cigarettes and other consumable items substituted for legal tender. Private German citizens knew that reforms must occur, but were uncertain of when. They also knew that the possessor of goods would, in the period after the reform, be able to sell them at good prices in a currency with value. Hoarding of goods became a common method of preparation for the currency reform, and this is one reason that the market could be flooded with previously unavailable goods immediately after it took place.

This much is common knowledge. But hoarding of goods and generalized expectations were not the only preparations made for the reform. A look at the inner workings of the Bayer factories in the months prior to June 1948 shows that preparations for the return to a money and market economy were specific and wide-ranging.

Internal discussions of the probable effects of the reform took place in Bayer sales meetings beginning in March 1948. Ulrich Haberland, the top executive at the former Lower Rhine Group, chaired the meetings, which took place every three to four weeks as needed. The heads of the firm's various departments—and especially the sales sections—attended, in all eleven to fourteen men. The impending currency revaluation formed an important part of the discussions in several meetings, but most notably in March and in early June 1948. In March, Bayer officials debated the reform's potential impact, especially on income from domestic sales. Bayer could anticipate a noticeable drop-off in the sales of dyes and pharmaceuticals in the aftermath of revaluation because of likely caches of reserves among consumers. On the other hand, since most customers buying bulk chemicals possessed only limited reserves, sales in this area should remain fairly good. A stagnation (*Stockung*) of the domestic market was, however, to be the likely initial impact of the reform. To bridge this over, the firm would have to depend upon forced export, which appeared especially promising in the area of organic intermediates. Company production units would prepare to take advantage of this projected demand for exports by building up stocks of Bayer products "for fulfilling spontaneous export possibilities."[16]

Discussions at the seventeenth sales meeting in Leverkusen on June 4, 1948, brought out still further the officials' concern and attempts to deal with the upcoming reform. Director Haberland had recently ordered a critical review (*Überprüfung*) of technical costs because of "the serious difficulties in markets and cash flow to be expected at least for a transition period with the enactment of the currency reform." Further discussion brought out these points:

The goal will have to be a reduction in our production costs, so that in the case of stagnation of the domestic market we can seek—and with competitive prices find—equalization in foreign markets. Sales sections will have to concern themselves with questions of expense. [The] chemicals [section] has already gone over to a system whereby sales costs are surveyed by section and through that will be truly capable of control. . . . The fixing of the exchange rate will be of major significance as well for further development.[17]

The timing of these discussions and their accuracy in terms of assessing potential problems arising from the reform were striking. Short-term shrinking of the market, the transition from raw materials to capital and credit shortage, reliance on export, and the long-term benefits of stabilization were only some of the issues that chemical industrialists anticipated. Bayer—but also other chemical companies and the chemical industry trade association—developed policies to try to manage the transition and to ward off its most threatening possibilities. The trade association sought to try to influence credit policy, for instance, while Bayer and other companies attempted to cut costs and regain footing in export markets.

Looking more closely at the cases of Bayer and Hoechst refines our picture of the actual effects of the reforms on western German industry as well. The immediate impact in Leverkusen was to place Bayer's liquidity in jeopardy. On July 2, Haberland reported to a sales meeting that the firm's obligations for the current month (including wages and salaries, purchase of raw and packing materials, and so on) came to about DM 55 million. The company had just DM 600,000 cash. Haberland went on: "the significance of ongoing sales, and the payments of our customers in cash or in discountable exchange is obvious enough." On the other hand, there appeared to be little cause for alarm. No production area had experienced a significant fall-off in sales, contrary to expectations. There was indeed a noticeable decline in sales of "staple products," but "the total picture [may be] termed favorable."[18]

The fact that the postreform situation at Leverkusen was slightly different from that expected beforehand does not invalidate the hypothesis that chemical firms benefited from discussions of and preparations for it. Prior preparation allowed the firm's leadership to manage the problems accompanying the reform rather than to be forced into unconsidered action. The Bayer sales committee, for instance, heard at the same July 2 meeting that the program to cut costs and expenses was already paying off, something that could only have a positive effect upon sales and liquidity. Paul Dencker reported that representatives from industry, the trade associations, and the banks in North Rhine Westphalia (NRW) had already met to determine industry credit needs owing to "the credit question, which has come strongly to the fore with the currency reform." The industry of NRW would probably need DM 1.5 billion in short-term credit, of which the banks could finance only one-half. Industry itself would have to help to finance the rest of the needs by putting up capital (*einen Haftkapital*) amounting to DM 150 million. The long-range plan was for industry to help found a Garantie- und Diskontbank A.G. to facilitate and finance industrial transactions. Former I.G. factories in NRW, the committee discussed further, would probably not have to rely on the bank for credit; in the interest of the provincial economy as a whole, however, participation was advisable. The discussion went on:

> Since at least the remainder of the chemical industry in its entirety will turn its eyes toward us as pacesetter, HABERLAND will consent after obtaining the agreement of [I.G. Control Officer] Mr. [Douglas] Fowles. He will add the proviso that our participation should secure us influence on the general granting of credit, and that, if possible, we should be represented by a member in the advisory body [*Beirat*].[19]

Later meetings of the firm's sales committee indicated that turnover slumped in the initial days after June 20, but within a month had generally reached normal levels once again. For the rest of 1948, production improvement continued across the board, especially in chemicals and dyes, although there were a few exceptions, including plant protective agents and pharmaceuticals. The major problem remained tightened liquidity, and it appeared likely by the end of July that short-term credit would have to be sought and a program enacted to cut daily cash outlays.[20]

Information on the Hoechst works is much more sketchy, but it is clear that their postreform experiences were similar. A "Survey of developments in the year 1948 at Farbwerke Hoechst" reported that "there were improvements in all fields after the currency reform." In fact exceptions to the rule existed—just as they did at

Leverkusen—in the areas of pharmaceuticals, plant protective agents, and insecticides. What is more, production suffered from inadequate supplies of dyestuff intermediates owing to the factory's dependence on former I.G. plants in other zones and in the east. This factor did not affect the more self-sufficient Bayer factories. Otherwise, production increases were steady and significant. Just as at Bayer, there were concerns at Hoechst over credit—for the factory itself and for its customers. Reductions in liquidity as a result of the currency reform also caused problems. The report noted:

> We were forced to postpone payments of wages and salaries for a few days at the beginning of July. . . . Furthermore it could not be avoided to make use of short term bank credits during the rest of the second half year.[21]

Attention to a single industry branch, and in particular to the experiences of key firms within it, helps refine our picture of the effects on western German industry of the reforms of mid-1948. Their impact on production and sales was indeed immediate, substantial, and in general positive. On the other hand, currency revaluation caused problems in the area of credit and liquidity. The beginnings of production recovery prior to the reforms, and the fact that problems accompanying them had to a large degree been anticipated and solutions to them sought in the months before June 1948, allowed the firms to take advantage of the positive aspects of the reforms and lessened their negative impacts.

But what was the effect on the fortunes of the chemical industry of the influx of Marshall Plan aid in mid-1948? First of all, aid from the European Recovery Program (ERP), the plan's formal name, was not the first—and not even the largest—U.S. program to aid the recovery of the western German economy. The Government Aid and Relief in Occupied Areas program, for instance, provided the German economy with large quantities of foodstuffs and petroleum products.[22] One must also look closely at the statistics related to Marshall Plan deliveries to western Germany (presented in table 11). Most ERP aid involved raw materials and foodstuffs, and what is more, that portion of the raw materials slated for the chemical industry was small. One may thus argue that the indirect effects of the ERP were important: foodstuffs allowed workers and

Table 11. ECA Deliveries to Federal Area and the West Sectors of Berlin from the Beginning of the Marshall Plan (to December 1951)

ECA Deliveries	In Thousands U.S. $	Percentage
Foodstuffs and agriculture	543,768	42.7
Industry	623,630	48.9
raw materials:	(549,999)	(43.2)
other:	(73,631)	(5.8)
Freight	107,058	8.4
Totals	1,274,456	100.0

Source: Adapted from table 1b in Bundesminister für den Marshallplan. *Neunter Bericht der Deutschen Bundesregierung über die Durchführung des Marshallplans* (Bonn, 1952), p. 112.

their families to be well fed; aid in the form of raw materials freed up German resources to buy other necessary goods, such as new machinery; and all of this was particularly crucial given the shortage of dollars after the war, especially in western Germany. To this extent, the ERP aided recovery. On the other hand, little of the materials obtained through the ERP were of direct use in the chemical industry. And it is certain from the magnitude of ERP aid as well as its timing that it did not *cause* the recovery or resurgence of the industry.

For the chemical industry—and indeed for all industry in western Germany—the Marshall Plan's importance lay less in its direct supply of goods in the short term than in its long-term supply of capital to industry. Goods shipped under the ERP program were either gifts or long-term loans from the U.S. government to the European governments. The American government paid for the goods, which were most often purchased from American sellers, in dollars. European state officials in return sold the goods in their respective countries for francs, pounds, marks, and so on. Proceeds from the sales went into special accounts, holdings which were known as counterpart funds. The accounts were then used to fund loans for industrial reconstruction, which had to be accompanied by a comprehensive and long-term program and approved by the U.S. government. In the case of West Germany, 1.27 billion dollars in counterpart funds were available by the end of 1951 to fund loans for reconstruction.[23]

For the chemical industry, the counterpart fund accounts were of immense value. In developing plans for investment for "Marshall Plan Year" 1949–1950, for instance, West German chemical manufacturers decided that they would need DM 190 million for reconstruction. DM 50 million would come from internal financing, while the remainder—DM 140 million—would come from loans from counterpart-funds accounts and from German financial institutions. The proportion of the DM 140 million financed by counterpart funds is not clear, but one may assume it was fairly large: for industry as a whole for the same year, DM 8.5 billion was to be invested, of which DM 2.5 billion, or just under 30 percent, would come from counterpart funds.[24] In sum, the Marshall Plan had a definite and positive effect on the recovery of the western German chemical industry. But the effect was most pronounced in the area of financing, and occurred much later than either the announcement of the program in June 1947 or the beginning of deliveries in April 1948.

INDUSTRY-SPECIFIC FACTORS IN THE MAINTENANCE OF RECOVERY IN MID-1948

The events just described were extremely important in explaining the development of the western German chemical industry, and in particular of the successors of the I.G., after mid-1948. They represented political and economic actions that had an immediate effect on the economy as a whole. Three further developments were of importance for the chemical industry in that they helped set the stage for its resurgence. They included the end of the trial of the former I.G. executives at Nuremberg, the founding of bizonal organizations to develop and implement policy with regard to the I.G., and the growing participation of the French in bizonal affairs. Let us examine each briefly.

The trial of twenty-three former I.G. executives began in Nuremberg on May 13, 1947. The defendants included Carl

Krauch, chairman of Farben's supervisory board; all eighteen members of the firm's managing board; and four lesser-ranking managers. Included in the last group were, for example, Heinrich Gattineau, former chief of the company's political economy office, and Walter Dürrfeld, the man who had managed the construction of I.G.'s Auschwitz plant. Prosecutors charged all twenty-three men with four crimes: preparing and waging a war of aggression; plundering and spoliation in German-occupied countries during the war; use of slave and forced labor and participation in mistreating and murdering the enslaved workers; and participation in a conspiracy to wage aggressive war. Points one and four were thus substantially the same, although point one accused the men of actually carrying out a crime against peace and point four accused them of conspiring to commit such a crime. A further count indicted three of the defendants—Christian Schneider, Heinrich Bütefisch, and Erich von der Heyde—for membership in a criminal organization, the SS.

The "I.G. Farben Trial" (Military Tribunal case VI) lasted more than a year and saw the presentation of thousands of documents by prosecution and defense. When the verdict came down on July 29 and 30, 1948, it startled and angered the prosecution. Ten of the twenty-three were acquitted of all counts. The court found all of the defendants not guilty of preparing and waging aggressive war and of conspiracy to wage a war of aggression; the charges of membership in a criminal organization were dismissed as well. Four of the remaining I.G. executives were found guilty on the slave-labor count (for their roles in starting and operating I.G. Auschwitz); eight were guilty of plunder and spoliation; and one man, Fritz ter Meer, was found guilty of both of these charges. Despite the gravity of the offenses, the thirteen guilty executives received relatively light sentences of between eighteen months to eight years imprisonment. Those implicated in I.G.'s activities at Auschwitz were punished with heavier sentences (six to eight years) than those who had plundered and despoiled foreign industry in German-occupied Europe.[25]

Why was the punishment so mild? Joseph Borkin, in his book-length indictment of the entire history of the firm (*The Crime and Punishment of I.G. Farben*), contended that the emergence of the Cold War between the United States and the Soviet Union influenced the majority of the court,[26] but it is difficult to imagine the precise mechanism through which this might have taken place. More readily apparent explanations are at hand, although we will never know for certain. First, the mild punishments fit with the American judicial tradition of light sentences for "white-collar crime." Use of slave labor and mass murder were, admittedly, not typical white-collar criminal activities; but, then again, some of the I.G. managers on trial had apparently known nothing of what was going on at Auschwitz, while it remained unproven that any of them had ordered or implemented actual extermination of enslaved workers.

A second potential explanation for the mild treatment of guilty I.G. executives is even more compelling: strategically, the prosecution conducted its case poorly at times. Prosecuting lawyers drew up an undifferentiated indictment; not all of the defendants *could* have committed all four major crimes they were charged with. Several of the defendants were uninvolved in the company's foreign activities and thus could have had at most an indirect role in plundering the European chemical industry. It was unlikely that lower-ranking I.G. executives had anything to do with an alleged conspiracy to wage aggressive war. What is more, the prosecution began its case on the wrong foot by introducing:

> organizational charts, cartel agreements, patent licenses, correspondence, production schedules, and corporate reports, as is done in antitrust cases, not at a trial of war criminals charged with mass murder.[27]

Borkin quoted U.S. prosecutors as saying that they realized early in the trial that they "should have started with Auschwitz," but by then the sequence of the trial had been determined by the order of indictments.[28]

I would like to suggest that, just as in the early treatment of the I.G. at the hands of American occupation officials (see chapter 2), the prosecution staff in the Farben trial was influenced—and partially entrapped—by its experiences with the firm during the 1930s. Josiah DuBois, the assistant prosecutor in charge of the case, had been associated with Bernard Bernstein and the Treasury Department during the war.[29] Other staff members had been in Treasury or the U.S. Department of Justice's Antitrust Division (including Borkin). They thought of Farben primarily as a trust and as a cartel organizer in the 1930s and continued to do so into the postwar period. These salient characteristics of the firm were, to them, the logical basis for its crimes and thus had to be presented to the court first. Furthermore, based on their earlier experiences of the I.G., they believed the firm was a monolith; consequently, all executives shared guilt for all crimes committed in the company's name. Their strategic mistakes in the case thus probably emerged naturally from a false sense that they "knew" the concern in its essentials before the trial began.

German writers, of course, saw the outcome of the trial differently. In this commentary, the deputy chairman of the chemical industry trade association for the Bizone (*Arbeitsgemeinschaft Chemische Industrie des Vereinigten Wirtschaftsgebietes*), Stefan Balke, pointed out that the implications of the trial extended much further than just to the individual fates of the men involved. He stressed that, "*The actual accused appears . . . [to be] the firm I.G. Farbenindustrie A.G.*" Because, furthermore, the I.G. had been an integral component of the German economy, and especially of its chemical sector, intimately tied to other firms and to German science and technology, "the accusations that the judgment against the I.G. bring up affect . . . the totality of the German chemical industry." Balke then drew his main point:

> The trial in Nuremberg has shown, that all organizational levels of our civilization—that is also the industrial economy—may be put in the position of having to make ethical and moral decisions on their own responsibility, especially in the case that in and of themselves apolitical technicians and businessmen find themselves in politically determined circumstances.[30]

At the same time, Balke noted that the I.G. had come into existence not to become a monopoly but rather "under the tough pressure of the facts of the world economy." He criticized the language used in the indictment against the defendants. And he quoted commentary on the sentencing from other German organizations to the effect that the defendants were "honorable men," and which proclaimed their "genuine humanity."[31] In other words, one had to take responsibility for one's actions, and even "apolitical" actors had to make moral and ethical decisions; but the I.G.'s executives, the company, and indeed the whole chemical industry should be excused from this because they were "honorable men" and had only done what economic reality had demanded. It was typical of the time that Balke did not take into account that five of the top managers of the most important German chemical corporation had been found guilty of charges related to use of slave labor and mass murder, and that there may therefore have been grounds for reexamining the fundamental values and attitudes of chemical industrialists.

But what were the implications of the trial's outcome for the German chemical industry and the I.G.? On the one hand, several of the defendants had been acquitted, and all had been acquitted on two of the five counts. The others were given relatively light jail sentences. Thus, some segments of public opinion could point out that the executives had received less punishment than they deserved. Others could say that the original indictments had been unjust in part, that those who deserved to be punished had been, and that it was time to get back to normal affairs. But probably the most important outcome of the trial from the standpoint of the recovery of the I.G. successors was the simple fact that it was over. No longer would world newspaper headlines carry news of the real and alleged crimes of the firm's employees and directors, and the successors could hope that the memory of such events on the part of potential customers would be short.

One of the successor firms, BASF at Ludwigshafen in the French zone, felt the effects of the trial's outcome much more concretely. Dr. Karl Wurster, former director of the Ludwigshafen–Oppau plant complex, was acquitted, and within weeks had returned to his old position.[32] His return meant that now two of the three directors of major I.G. factory complexes in the western German areas from the Nazi period were back in office (or, in the case of Haberland, had never left it). Wurster's resumption of his former position also fit in well with the French policy of maintenance of the former I.G. works group virtually intact, and of attempting to rebuild production at the former BASF.

Wurster was the only defendant at the I.G. Nuremberg trial to play a prominent and active role in the management of a major successor corporation to the I.G. But many of the other defendants—whether acquitted or found guilty—also took up positions in the chemical industry after 1948. Fritz ter Meer, for instance, who had been found guilty at the trial of plunder and spoliation and slavery and mass murder and sentenced to seven years in prison, became chairman of the supervisory board of Bayer A.G. in the mid-1950s. Heinrich Gattineau, who had played a part in the famous 1932 meeting between I.G. representatives and Hitler (see chapter 1), but who was acquitted of the charges against him at Nuremberg, later became a member of the managing board of WASAG, a chemical company formerly associated with the I.G.[33] Others found themselves closed out from direct participation in the industry but sometimes managed to maintain contact with (and perhaps influence upon) their former colleagues.[34] The leadership of the German chemical industry apparently changed but little after World War II,[35] probably aiding rapid reconstruction and resurgence but leaving open the question of the extent to which German society had been reformed.[36]

During the summer of 1948, a second major development took place that was important for the Farben successors: the founding of bizonal organizations to conduct policy on the I.G. The formation of such organizations was a response to the failure of four-power attempts to deal with the firm, as the directive creating the organizations made clear:

> Since as a consequence of the breaking off of the meetings of the four power committee that was formed on the basis of Control Council Law Nr. 9 further progress in the break-up of the property of the former I.G. Farben concern has become impossible, it is seen as desirable to set up joint organizations for the implementation of the division of the assets of the I.G. Farben in the Combined Economic Zone [Bizone].[37]

The directive created two organizations. The first, the Bipartite I.G. Farben Control Office, or BIFCO, was composed of British and American I.G. Farben control officers and their representatives and was subordinated to the Bipartite Decartelization Commission. BIFCO had primarily a watch-dog function; it was responsible for supervising the work of the German organization also created by the directive.

FARDIP, or the I.G. Farben Dispersal Panel, was the parallel German organization to BIFCO. The bizonal Administrative Council (*Verwaltungsrat*) chose a minimum of six and a maximum of eight members for the panel, although the two chairmen of BIFCO had veto power over the appointments. FARDIP's powers were far-reaching. It took over supervision and administration of the I.G. property in the bizonal area from Allied bureaucrats and assumed responsibility for ending the company's cartel relationships and for drawing up plans for its breakup.[38] Creation of the organization thus marked a major turning point in Anglo-American policy toward the I.G.: although Germans had advised the Allies on policy toward the concern throughout the occupation period, officials of the defeated nation obtained for the first time a formal and substantial say in bizonal policy toward the company.

But who were the Germans who would have a say in the breakup and administration of the I.G.? The composition of FARDIP is telling. Six men made up the panel, the most prominent of whom was Geheimrat Dr. Hermann Bücher, who had been active in German politics and the economy since the First World War. In the early 1920s, Bücher helped found the country-wide umbrella organization *Reichsverband der deutschen Industrie*, or Association for German Industry. Through that connection he came to know I.G. Farben's Carl Duisberg quite well. By the middle of the decade, Bücher became an economic adviser to the chemical combine and a member of its economic subcommittee. He was, however, best known for his connection to German electrical industry, being named to the managing board of AEG (*Allgemeine Elektizitätsgesellschaft*, or German General Electric) in 1928 and becoming the firm's general director in 1933. Bücher had been active in the German economy during the Nazi period, and indeed was "reputed to have had a close friendship with Goering and subsequently with Speer." He was thus "in the automatic arrest category of US Military Government under instructions issued in 1945."[39]

Geheimrat Regierungsrat Dr. Gustav Brecht was another prominent member of the panel. Brecht, too, had been closely associated with government during the Nazi years, representing in particular the interests of the Rhenish brown coal industry. He also had previous contact with I.G. Farben: "As Deputy Chairman of Braunkohlen- und Benzin A.G. [BRABAG] he is reputed to have been closely associated with Dr. Bütefisch of the Farben organization." (Bütefisch, a member of the I.G. managing board, was especially active in synthetic fuel production.) Brecht was also on the U.S. military government's automatic arrest category in 1945. One further member of FARDIP, Freiherr Egon von Ritter, had been in the same category. He had been a member of the Nazi party (although he was denazified after the war) and allegedly helped the I.G. acquire control of the German magnesite industry during the Nazi period by using his position and influence as a banker.[40]

The remaining panel members were less well-known but no less important. Dr. Oskar Loehr and Dr. Eugen Moehn had both been employees of the I.G. Moehn was a lower-ranking chemist "who reputedly was placed on FARDIP as a representative of the Trade Unions." Loehr, on the other hand, had been an I.G. chemist and was after 1945 assistant director at Leverkusen. The U.S. I.G. Farben control officer's report on the composition of the panel noted that "confidential information, believed to be reliable, is to the effect that he was nominated to FARDIP through the activity of former Farben Vorstand (managing board) members and of Dr. Haberland, General Director of the Leverkusen complex."[41]

As the only FARDIP member with no ties with the I.G., Dr. Arnold Burghartz was formerly involved with the supreme court in Cologne, and it was "reported that he was nominated for FARDIP over the opposition of Farben interest [sic] and Farben representatives." The control officer's report continued:

> Independent investigation considered reliable indicates that he is reported to have been a resister to the Nazi Party and its program with consequent personal loss of position necessitating his retiring from the active practice of law.
>
> He has the further reputation of being sincere and dependent [sic] and of being an individual of high standing in his profession.[42]

The body responsible for developing and implementating policy on the I.G. Farben was thus dominated by men sympathetic to the object of their investigation. Three of the six had been on intimate terms with members of the I.G. managing board, and with high members of the Nazi regime. Two further FARDIP members had worked for the firm, and one of them—Dr. Loehr—continued to work for a successor.[43] Only one person on the panel was independent of Farben. The staff of the FARDIP committee reflected even more the tendency to employ personnel formerly associated with the I.G. Paul Dencker, for instance, was formerly the chief accountant (*Chefbuchhalter*) of the I.G. and a member of the firm's commercial and technical committees. Later a member of the managing board of Bayer A.G., Dencker was an important adviser to FARDIP.[44] The panel's successor organization, given the task of implementing its suggestions, was the I.G. Farben Liquidation Committee. The liquidators were prominent persons without direct connection to the I.G. But, more to the point, their advisers were personnel formerly associated with Farben.[45]

Establishment of these organizations in summer 1948 was important since German nationals obtained a much larger say in issues connected with the company. What is more, the Germans who had the largest say were ones who represented and/or sympathized with the interests of the concern. The western Allies thus by mid-1948 ran into one of the common obstacles to reform of industrial structure in a complex economy. Reform proved extremely difficult without consultation with representatives of the firms to be restructured. Consultation with such men had the effect of surrendering to them some measure of power over how the reform would take place.

Growing participation of the French in western Allied policymaking and practice was the final major development in mid-1948 of importance to the chemical industry, and in particular to the I.G. successor firms. The very fact that the currency reform applied not just to the Bizone but to all of the western zones, provided momentum toward policy unity. Unified currency revaluation also facilitated the transfer of goods between the Bizone and the French zone. On the other hand, the French military government at first refused to change the Bizone into a Trizone. BIFCO, for instance, the bipartite agency responsible for supervising FARDIP, became TRIFCO (Tripartite I.G. Farben Control Office) only in the latter part of 1949. And, although representatives from the western zones met with representatives of other western European nations in the OEEC (Organization for European Economic Cooperation) to decide on how best to divvy up Marshall Plan aid, there were separate contingents from the Bizone and from the French zone well into 1949.

Limited progress toward creating tripartite government in occupied western Germany did not preclude preparation for the day when a unified West Germany would come into existence. By April 1949, for instance, the United States' I.G. Farben control officer, Myron Maupin, was able to report the conclusion of the seventh "unofficial meeting" between the control officers of all three western zones. The number of meetings that had already taken place indicated growing communication between (if not necessarily the policy coordination of) western military governments on I.G. Farben. What is more, Maupin reported that the unofficial meeting of April 7 marked an "important turn, in that the meeting concerned itself largely with future activities between the Bizone and the French Zone either in a TRIFCO set-up or as a coordinated effort." The French control officer agreed in principle with the approved instructions to FARDIP, and the three officers planned examination of, and action upon, issues related to the I.G. The control officers themselves—in conjunction with a representative of the French control officer working with the members of FARDIP—would coordinate such examination and action.[46]

THE LONG-TERM EFFECTS OF THE EVENTS OF MID-1948

The six developments described above occurred within a relatively brief span of time during the summer of 1948. Because they together helped bring about and sustain a major jump in chemical production, all were important. The longer-term impact of developments in mid-1948 on the former I.G. firms is apparent from table 12. The year 1948 was a relatively good period for all three plants: turnover increased over its level in 1947 by more than 60 percent at Bayer, at BASF by more than 35 percent (this despite a devastating explosion at the Ludwigshafen factory in July 1948), and at Hoechst by one-half. These are substantial improvements in production. On the other hand, they are not particularly spectacular in terms of those of the early federal period. Turnover did not grow much from 1948 to 1949, except at Leverkusen, where the increase was 22 percent. But from 1949 to 1950, turnover jumped at Bayer by 40 percent, at Hoechst by over 50 percent, and at Ludwigshafen by over 80 percent (partly as a result of the repairs following the explosion)! The same is true for the period from 1950

Table 12. Turnover of the I.G. Farben Successors, 1946–1954 (million RM/DM)

	BASF[a]	Bayer	Hoechst[b]
1946	132	188	80
1947	188	216	99
1948	260	348	148
1949	261.7	424.4	164
1950	482.3	599.4	252
1951	678.6	923.8	340
1952	661.8	866.9	n.a.
1953	885.2	1025.1	942.6
1954	1050.0	1209.6	1126.7

[a] Does not include turnover from subsidiaries.
[b] Includes turnover from subsidiaries with greater than 50-percent stock participation by Hoechst.
Source: Compiled from estimates from bar graph of "Gesamtumsätze" in Bayerwerksarchiv, Leverkusen, Finanzwesen: Gesamtumsätze, 15.D.2; and from figures presented in Alfons Metzner, *Die chemische Industrie der Welt*, Vol. I: *Europa* (Düsseldorf: Econ Verlag, 1955), pp. 139, 147, 165.

to 1951, when turnover increased by 55 percent at Bayer, by 40 percent at BASF, and by 35 percent at Hoechst.

In summary, the currency reform and accompanying liberalization of the economy helped to eliminate the black market and the so-called "*Kompensationsgeschäfte*," or industrial barter market. They also helped create a stable environment in which businessmen could operate effectively. The end of the I.G. Farben trial allowed an important wartime figure, Carl Wurster, to return to his post as head of the BASF factory complex. More importantly for the chemical industry as a whole, and for the other I.G. successors, the end of the trial permitted progress toward wrapping up the breakup of the firm. It allowed hope that the bad publicity the trials involved would lose its effect on former and future customers.

The beginning of formal German input into Allied policy on the I.G. through the formation of FARDIP had significant implications. Actual control over planning and administration of the former I.G. property by Germans involved decisive day-to-day influence over the breakup. The fact that sympathizers of the I.G. dominated FARDIP meant that control would be used to further the interests of the former chemical giant. Increasing participation of the French in bizonal policymaking allowed the first moves toward a west German solution to the problems associated with the breakup of the I.G. Finally, aid under the Marshall Plan was important, although the primary impact of such aid was felt much later in the area of credit available for reconstruction.

By 1950, as production levels and export levels in the industry began to approach 1936 ("normal") levels, one could say with certainty that recovery was underway and, furthermore, that recovery dated from mid-1948. Still it is important to keep in mind that there were qualifications on this. First, there were areas of continuity in economic recovery. The chemical industry—and indeed all industry in western Germany—had begun to recover by 1946, and the factors noted had an influence only because the material basis for recovery was there: that is to say, industrial plant remained surprisingly intact after the war, and, for the I.G. successors in particular, key personnel often remained in places of influence and responsibility within the firm. Anticipation of the reforms and preparation for them were also important elements of continuity from the prereform to the postreform period. Throughout the occupation, Germans continued to exercise some influence on policy, especially through the state apparatus in the U.S. zone and through state and industry in the British zone. Such influence could more easily be translated into action after mid-1948, of course. Finally, there was a trend prior to mid-1948 toward unity of Allied policy, dating indeed from before the official founding of the Bizone.

The second qualification on seeing mid-1948 as an absolute breaking point has to do with what occurred after that eventful summer. Significant problems continued to exist for all industry after the currency reform. We have noted some of them already. For BASF, for instance, the immediate post-currency-reform period involved major destruction to plant and severe production cutbacks after an explosion rocked the factory in July 1948.[47] Other factories faced problems of liquidity and shortage of credit. All of the former I.G. factories labored under the uncertainty over what would be the outcome of the ongoing breakup of the firm. Coal supply became less problematic but still loomed large as a factor limiting production.[48]

Recovery of the western German chemical industry, and especially of the I.G. Farben component of it, was therefore well underway by 1950. Resurgence remained in the distance. When the Korean War came in 1950, it was at first a mixed blessing. Production to fulfill renewed demand owing to military spending promised to bring industry throughout the western world out of an impending slump. On the other hand, production increases placed enormous stress on the weak link in the west German—and indeed the western European—recovery, the coal industry. In the short term, the result was an energy crisis.[49] Overcoming that crisis and laying the bases for the resurgence of the chemical industry are the subjects of the next chapter.

7
NEW BEGINNINGS AND RESURGENCE 1950–1951

Especially after mid-1948, the gradually emerging results of changed occupation policies, economic reform, foreign aid, and increased German influence led to remarkable recovery in the western German chemical industry. Dr. Ulrich Haberland, the director of the Leverkusen complex, in early 1951 noted: "In 1950 West German chemical production has in general reached the level of 1936 once again, in some areas even surpassed it."[1] Important problems served, however, to constrain the industry from growing at an even faster rate. Hoechst, BASF, and Bayer overcame the difficulties as a result of key developments during 1950 and 1951. They included major progress in the legal breakup of the I.G. and in the establishment of successor companies, the resurgence of export, procurement of investment of capital, and resolution of trade mark disputes. But before looking at them, it will be useful to examine the dimensions of the production improvements in the early 1950s that constituted the resurgence of the I.G. successor companies.

THE DIMENSIONS OF RESURGENCE: PRODUCTION IMPROVEMENT AND THE BEGINNINGS OF RESURGENCE IN THE EARLY 1950s

At least symbolically, 1950 marked a turning point for the German chemical industry, as well as for many other branches of the German economy, as the figures in table 13 indicate. For the first time, chemical production exceeded its prewar output in the area of the Federal Republic. Later in the decade, the industry experienced

Table 13. Development of Production for Selected Production Areas in West Germany, 1948–1954 (1936 = 100)

Calendar/Monthly	Coal Mining	Iron & Steel Products	Machine Tools	Chemicals Industry	Textiles
Weight	6.66	5.34	8.05	8.66	7.49
1948	78.5	37.7	55.7	69.7	51.0
1949	92.4	63.0	97.1	96.1	89.5
1950	98.8	81.7	123.7	125.3	118.6
1951	107.5	92.9	164.5	150.1	130.2
1952	111.7	107.3	188.7	154.7	125.2
1953	112.9	102.2	184.5	182.0	145.2
1954	115.2	114.5	208.1	210.4	150.8
1953 Jan.	119.1	120.1	167.4	166.2	139.5
Feb.	107.6	105.8	171.2	161.5	124.0
Mar.	118.9	115.0	197.8	180.0	141.5
Apr.	107.1	99.9	180.5	177.6	135.8
May	102.8	94.7	176.1	170.9	124.0
Jun.	110.7	101.4	191.4	177.9	138.6
Jul.	116.7	102.2	184.1	185.8	146.2
Aug.	112.6	96.2	171.8	184.7	144.3
Sep.	113.6	96.9	200.2	193.9	166.3
Oct.	120.1	100.0	193.5	201.9	174.7
Nov.	112.3	97.1	181.5	194.2	158.8
Dec.	113.5	96.6	199.0	190.8	149.2
1954 Jan.	116.5	98.1	173.2	187.8	142.8
Feb.	111.6	94.9	182.2	184.8	141.6
Mar.	122.7	108.2	208.1	212.1	157.6
Apr.	106.5	98.1	195.0	201.5	139.6
May	108.9	106.9	203.6	208.3	139.8
Jun.	106.3	108.6	203.5	205.1	131.7
Jul.	119.8	125.3	213.2	216.0	148.9
Aug.	115.6	120.8	196.3	216.3	148.9
Sep.	116.4	123.6	224.7	224.5	168.9
Oct.	119.9	128.5	223.9	225.9	167.9
Nov.	117.9	131.6	225.2	220.8	163.3
Dec.	120.5	129.7	248.8	222.0	158.8

Source: Adapted from *Statistisches Jahrbuch für die Bundesrepublik Deutschland 1955*, ed. by the Statistisches Bundesamt (Stuttgart: W. Kohlhammer, 1955), pp. 228–230.

Table 14. Development of Production in the Chemical Industry in Important Countries, 1948–1958 (1938 = 100)

Year	F. R. Germany	Belgium	France	Great Britain
1948	54	144	113	164
1949	78	146	116	182
1950	104	160	121	213
1951	125	208	148	228
1952	127	190	137	215
1953	149	192	149	256
1954	170	225	170	282
1955	193	229	197	297
1956	210	248	213	310
1957	234	246	234	321
1958	234	248	264	315

Year	Italy	Netherlands	Norway	Austria	Canada	USA
1948	109	107	130	145[a]	n.a.	252
1949	117	128	160	168	n.a.	241
1950	140	155	213	206	217	286
1951	183	165	255	213	243	321
1952	172	155	240	196	254	324
1953	213	167	250	213	286	345
1954	262	183	300	249	309	345
1955	187	195	315	268	343	386
1956	317	207	315	274	371	407
1957	334	222	323	302	386	417
1958	377	227	295	311	391	414

[a] Figures for Austria on the basis of 1937.
Source: *Chemiewirtschaft in Zahlen* (Zusammengestellt vom Verband der chemischen Industrie, e.V.), 4th ed. (Düsseldorf: Econ, 1960), pp. 90–91.

spectacular growth: production nearly tripled at the same time that industry as a whole doubled its production rate.[2]

Table 14 shows that the experience of the Federal Republic's chemical industry was by no means unique. (Note that here 1938 = 100.) All industrial nations, in fact, shared in the boom.[3] Production in France more than doubled between 1950 and 1958, while Italian chemical production rates rose by two-and-one-half times during the same period. Even the chemical industry in the United States, which was already producing at nearly three times its 1938 rate in 1950 owing to wartime expansion, limited destruction, and continued production booms in the immediate postwar period, increased output by nearly one-third between 1950 and 1958. Two factors differentiate the performance of the West German chemical industry from that of all others, however. First, pre–World War II German industry manufactured chemicals at a level far exceeding that of most other countries. Second, 1938 was already a boom year for that industry, since investment under the Four Year Plan and preparations for war were in full swing. Thus it may be said that, from the point of view of production, the West German chemical industry experienced a resurgence during the 1950s that far exceeded the general recovery of the immediate postwar years. Parallel figures for the production performance of the I.G. successors during the 1950s are not available. But we can surmise from the trend in their turnover during the decade that they also experienced a resurgence (see fig. 8).

Development of export for the chemical industry was a major driving force behind the production boom (see table 15.) In 1937, for instance, the chemical industry was responsible for 15 percent of German export by value. During the immediate postwar years, however, it did not participate in the export economy to as great an extent. Chemical exports were negligible from 1945 to 1947 and amounted to only 7 to 8 percent of total sales abroad in 1948 and 1949. It was not until 1950 that chemical exports first approached "normal" levels, when they rose to 13 percent. What is more, industry exports remained at the level of 12.5 to 14.5 percent of total export proceeds from the Federal Republic of Germany (with one exception, in 1952) even as total sales abroad increased from DM 8.362 billion in 1950 to DM 41.184 billion in 1959. In other words, chemical producers kept pace as West German exports increased in value by nearly five times in nine years!

Exports drove the resurgence of the I.G. successors as well during the 1950s (see table 16). Once again, export revenues as a percentage of total turnover lagged in the immediate postwar period for all three of the successors. By 1951, however, export proceeds accounted for approximately one-third of total turnover for each of the three. They remained around this level throughout the 1950s, and sales abroad thus kept pace with the general sales boom of the decade.

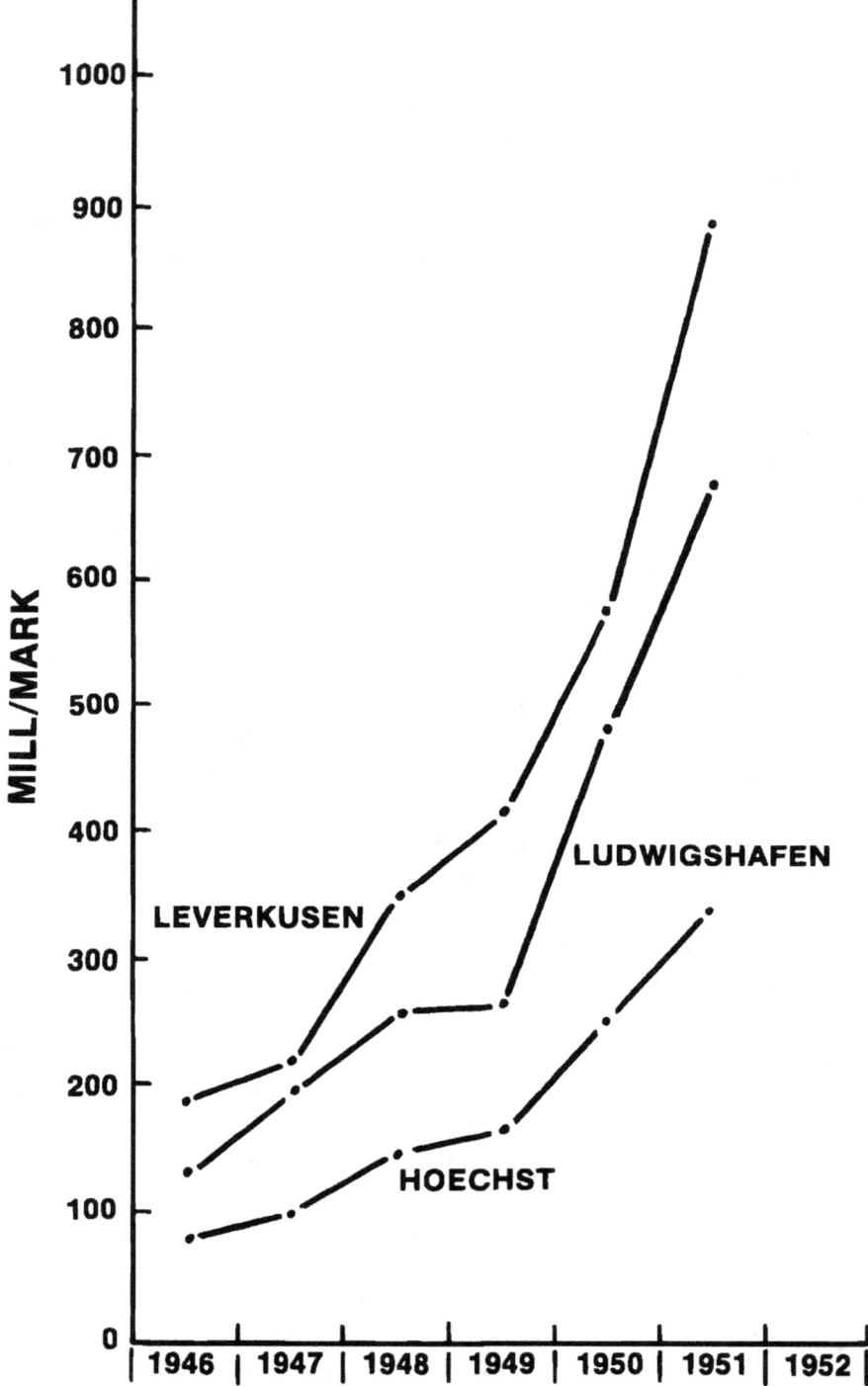

Figure 8: Turnover of the I.G. Successor Companies, 1946–1951.

Source: Adapted from graph (n.d.) in Bayerwerksarchiv, Leverkusen, Finanzwesen Gesamtumsätze, 15.D.2.

Table 15. Chemical Exports from the Federal Republic of Germany, 1937–1959 (in 1000 RM/DM)

Year	Chemical Export	Total Export	Share of Chemical Export in Total
1937	889,076	5,911,000	15.0%
1948	125,418	1,816,891	6.9
1949	312,785	3,805,514	8.2
1950	1,083,649	8,362,134	13.0
1951	2,113,913	14,579,428	14.5
1952	1,772,538	16,908,834	10.5
1953	2,323,979	18,525,579	12.5
1954	2,957,927	22,035,206	13.4
1955	3,396,913	25,716,788	13.2
1956	3,906,786	30,861,036	12.7
1957	4,503,766	35,968,043	12.5
1958	4,623,083	36,998,055	12.5
1959	5,449,718	41,183,907	13.2

Sources: Adapted from Arbeitsgemeinschaft der chemischen Industrie, *Tätigkeitsbericht*, 1950, pp. 31–32; ibid., 1951, pp. 41–42; *Chemiewirtschaft in Zahlen*, 4th ed. (Düsseldorf: Econ, 1960), p. 29.

Table 16. Export of the I.G. Successors, 1946–1951 (as a percentage of turnover)

Year	Leverkusen	Ludwigshafen	Hoechst
1946	0%	18.2%	0%
1947	3.8	16.7	4.0
1948	9.1	18.5	8.0
1949	15.0	19.2	12.2
1950	26.4	29.3	27.0
1951	33.3	32.9	32.9

Source: Calculated from figures in bar graph of "Gesamt-Umsätze," n.d., Bayerwerksarchiv, Leverkusen, Finanzwesen Gesamtumsätze, 15.D.2.

THE BASES OF RESURGENCE: POWER POLITICS AND THE WESTERN GERMAN STATE

In order to come up with an explanation for the statistics showing the production increases in the early 1950s, it is necessary to examine a number of qualitative changes both in the industry and in the context within which the industry operated. The most important factors included the role of power politics and the growth in official power of the western Germans, the manner in which the disentanglement of the I.G. took place, the founding of the successor firms and their leadership, the sources of investment for the new firms, and the integration of the firms into the U.S. postwar economic order. Let us examine each in turn.

From the early occupation period, the trend in western Germany was toward returning some measure of power to the Germans. The Americans were the first to make moves in this direction, mainly out of the desire to keep the occupation as short and as inexpensive as possible. Germans in the American zone helped advise upon and implement decisions while the small zonal military government apparatus supervised their actions.[4] The tendency continued into the bizonal period, as the United States gained more and more influence over the conduct of the occupation. In fact, a chemical industry official in the French zone of occupation, writing in mid-1949, commented sourly that Anglo-American control of the chemical industry amounted only to "the examination of statistics."[5] By the time the Americans and British combined zones in early 1947, furthermore, the incipient conflict between east and west known as the Cold War engendered a major change in U.S.—and later in western Allied—policy. The Allies came to look upon western Germany not as a defeated enemy state to be held down, but rather as an ally to be built up politically and economically as a bulwark against communism. Consequently, even more control was turned over to the Germans.

As the first superzonal German governmental organization, the Bizone served as the basis for the future West German government.[6] Men active in the German bizonal administration later held similar offices in the federal period. Events in 1948 hastened the creation of a West German political entity. The currency reform of June 1948 connected all three of the western zones for the first time since 1945, at least on the financial level. Other areas of cooperation followed. The Soviet military regime in the eastern zone of Germany responded to the currency reform within days. On June 24, the Russians completely shut down access to the western zones of Berlin. The blockade lasted until May of the following year and the separate currency reform in the eastern zone meant that there were now, for all practical purposes, two Germanies. Monetary unification in the three western zones on the one hand, and their separation from the eastern zone on the other, paved the way for the creation of the West German state. The western Allies agreed through their representatives on April 8, 1949, to an Occupation Statute defining the conditions under which that state was to be created, and one month later, on the eighth of May, the Federal Republic of Germany came into existence with its capital in Bonn, under Chancellor Konrad Adenauer.

The change in governmental structure in the western German area involved a change as well in Allied military government. The three separate occupation units were trimmed drastically; in most areas, an Allied High Commission took their place. The change also meant a reduction in Allied personnel involved in the control and direction of Germany, as well as in the power of remaining personnel. Still, the high commissioners and their subordinates retained control over several vital areas, including administration of the Ruhr area, dismantling of "surplus industrial capacity", and control of war-related industries. For the chemical industry, Allied power continued to have daily and decisive implications. Ultimate control over the disentanglement of the I.G., for instance, remained in the hands of the representatives of the high commissioners. The western Allies also set up a Military Security Board (MSB) with extensive powers and responsibilities in the area of enforcing restrictions on chemical production in the western German area. The MSB seems to have been an expression of continued Allied fears of a resurgent—and dangerous—German chemical industry.[7]

Even before the creation of the West German state, Germans had maneuvered to gain more power over their own affairs. Under Chancellor Adenauer, the push toward sovereignty took on new dimensions. Adenauer took advantage of the Allied, and especially the American, desire to form a strong and loyal West German state to gain concessions in several areas. Surrender of some control to the Germans under the Occupation Statute involved, to be sure, by no means a complete turnover of authority to the new government. In fact, on the very day the statute was enacted (April 8, 1949), the foreign ministers of the United States, United Kingdom, and France signed the Washington Agreement on Prohibited and Limited Industries in Germany.[8] Among other things, the agreement restarted dismantling in certain critical areas—such as synthetic fuels production—and restated some Allied production prohibitions. Particularly hard hit by the policy were the synthetic facilities in Germany's Ruhr area.[9] Resumption of some of the most hated occupation practices at precisely the moment that the West German state was coming into existence aroused a major outcry in the German public, and from the new government.[10]

The outcry and redoubled pressure for change from the new West German government bore fruit rapidly. On November 22, 1949, Adenauer and the three Allied high commissioners signed the Petersberg Agreement, named for the hill overlooking the new German capital of Bonn upon which the high commissioners had their offices. Thereafter, the Federal Republic of Germany could participate as an independent actor with other western European states in making decisions on European recovery (of particular relevance with regard to the OEEC). What is more, seven iron works and eleven chemical firms (including the synthetic fuels plants) in the Ruhr were removed from the reparations and dismantling list. In return, the federal govenment accepted the creation of the International Authority for the Ruhr[11] and received a seat on that authority. In other words, in return for agreeing to give up some sovereignty over the Ruhr area, the new regime received recognition as an independent state within Europe in most other areas and major concessions on the dismantling/reparations front.[12] Since, moreover, the International Authority proved to be short-lived, the West German govenment eventually emerged with its sovereignty fairly intact with regard to policy on the Ruhr. The offer on May 9, 1950, of French foreign minister Robert Schuman to create a European Coal and Steel Community with authority over the coal and steel industries of all of the member states— including among others France and Germany—placed Gemany on equal footing with other western European states.[13]

The fact that the Petersberg agreement eliminated the problem of dismantling for a number of firms did not mean that all controls over German industry were removed. Production prohibitions in the areas of synthetic fuels, synthetic rubber, various other chemical products (including chlorine, caustic soda, and other important basic chemicals), and several nonchemcials, remained in place. Industry and the federal government continued to agitate for removal of such prohibitions.[14] Again protests paid off, in part because of the need of the Allies—in particular the United States—for German support in the Korean conflict and for the beginnings of German agreement to rearmament. On April 3, 1951, the high commissioners and the federal government agreed to lift prohibitions of synthetic fuel and rubber production (among others). The commissioners also loosened restrictions in several other production areas.[15]

The German government, supported by chemical producers, thus maneuvered itself into a position of almost full sovereignty by 1951, although suspension of remaining Allied controls did not occur until 1955. The chemical industry in particular was now able to produce almost anything it pleased. The process whereby the federal govenment came into being and regained sovereignty thus is crucial to understanding the preconditions for resurgence of the German chemical manufacturing. The struggle for sovereignty remained to be fought in some critical areas, however, among them the breakup of I.G. Farbenindustrie A.G.

THE BASES OF RESURGENCE: THE "ENTFLECHTUNG" OF THE I.G.[16]

With the end of the war in April/May 1945, I.G. Farbenindustrie A.G. was effectively broken up, *de facto* if not *de jure*. Each of the four zones of occupation was sealed off from the others, economically, politically, and socially. Formal confirmation of the breakup followed. The Americans were first to act in early July 1945 when they declared that property of the firm in their zone would be confiscated and administered by military government personnel and that the I.G. would be dismantled into smaller units. A similar French law took effect two weeks later. The Russians meanwhile took over control of the I.G. factories in their zone, while the British waited until mid-November to pass their own law to seize the assets of the firm. The separate Allied actions were ratified at the four-power level on November 30, 1945, with the enactment of Allied Control Council Law No. 9.

Four-power attempts to devise a mutually satisfactory solution to the problem of the breakup of Farben continued into 1948, but were never very effective. Policy action (or often inaction) at the level of the individual zones was much more important in the early years of the occupation, and it differed substantially from zone to zone. The Soviets, for instance, developed a common policy for all of the I.G. factories in their zone. They ordered production to begin again in the enormous Leuna Works of the former I.G. by July 21, 1945,[17] and seized Leuna and several other large factories in the name of the Soviet state. The imposing of reparations on, and the heavy dismantling of, I.G. factories in the Soviet zone limited production and efficiency, however. The French and the British also treated the major I.G. complexes in their zones as single units, although for different reasons. The French kept the two major factories of the former BASF in their zone together in order to seize their production more effectively. The British allowed production relationships among the four factories of the Lower Rhine Group to continue after the war, going so far as to retain in his previous position the wartime manager of the group, Ulrich Haberland. Their concern was with the decision over nationalization or continued private ownership of private firms. In either case, the Bayer factories would remain a large, integrated production unit. American policy toward the I.G. differed from all of the others. They were the first to seize the firm's assets, and they adopted the most radical measures to break it up. All of the top managers of the firm were fired in the American zone. The factories of the former Maingau Group had to operate independently of one another; some members of U.S. military government went so far as to suggest a breakup of the main Hoechst *factory* itself into several independent companies.[18]

All semblance of four-power cooperation broke down by the spring of 1948. This failure paved the way for more effective and coordinated policy toward the I.G. above the zonal level. The British and Americans created in August 1948 the Bizonal I.G. Farben Control Office (BIFCO) and the parallel German organization, the

I.G. Farben Disperal Panel (FARDIP), which were responsible for developing plans for the final disposition of the chemical giant. Anglo-American organizations provided some measure of policy unity in two of the four zones. The emergence of an effective German organization concerned with the problems of the I.G. breakup marked a new stage in the disentanglement process. Germans were thereafter to have an increasing say in how the firm was to be broken up.[19]

The year 1950 was a key time in the disentanglement process for several reasons. On June 29, FARDIP delivered its plan for the breakup of the concern to the western Allies and then disbanded. The plan was notable above all for introducing the notion of *Kerngesellschaften*, or "nuclear firms," into the discussion. Representatives of FARDIP, motivated primarily by the recognition of Germany's need to export in order to survive in the postwar world and by an awareness of the place of the chemical industry in that effort, called for the creation of three firms of approximately equal size centered around the old works groups of the I.G. Although the FARDIP plan called for twenty-two independent units in all, Bayer, BASF, and Hoechst were to dominate the western German chemical industry and to be of sufficient size to compete effectively on world markets.[20] FARDIP's conception of *Kerngesellschaften* became the heart of German policy with regard to the breakup of the I.G., and eventually was accepted even by the Allies.

The year 1950 proved pivotal for other reasons as well. On August 17, Allied High Commission Law No. 35 provided the legal basis for the stock transfers that would break up the firm. To carry out the law more effectively, the combined French, British, and U.S. group responsible for the breakup of the firm became officially TRIFCOG, or the Tripartite I.G. Farben Control Group. The trend toward organizational (if not always policy) unity among the western Allies was thus completed.

Implementation of Law 35 also required more formal German input. Chancellor Adenauer and Economics Minister Ludwig Erhard continued to take an active personal interest in the disentanglement, but the task of coordinating and developing German governmental policy was turned over to two men within the Economics Ministry, Dr. Felix Prentzel and Dr. Heinrich von Rospatt.[21] Both Prentzel and von Rospatt were former employees of the I.G.

who had moved into government after being laid off by the Allies; Prentzel in particular had been active in the government's handling of I.G. matters since at least 1948.[22] They recruited several advisers, including the prominent banker Hermann J. Abs, the chemical industrialist Helmuth Wohlthat, chemical industry trade association president W. A. Menne, and former FARDIP member Gustav Brecht. Ministry personnel and their advisers, meeting at times every day during late 1950 and 1951, also drew representatives of the former I.G. works groups into their discussions. Prentzel was responsible for organizing the first official postwar meeting of the heads of the groups in January 1950,[23] and industrialists provided vital input into the negotiations throughout the disentanglement process.[24] Trade union officials supported these efforts of government and industry. Prentzel and von Rospatt, representing the Federal Ministry of Economics in meetings with TRIFCOG, could therefore present German arguments confident that their views were supported by all major groups affected by the disentanglement.[25]

Toward the end of the year, on November 23, 1950, a committee of experts appointed by the Allies presented its findings on the problem of the I.G.[26] Erwin H. Amick (an American chemist), George Brearley (a chemical industrialist from Great Britain), and the French chemist Leon C. Denivelle accepted the FARDIP idea of *Kerngesellschaften*, which involved a major concession to German wishes.[27] There were significant differences between the Allied recommendations and those of FARDIP, though: according to the Allied experts, Dormagen, for instance, was to be separated from the Lower Rhine (Bayer) Group and joined with the former I.G. plants in Bobingen and Rottweil into a company specializing in synthetic fibers; the Agfa factories at Leverkusen and Munich were to be independent; and the Casella works at Mainkur were to be a separate company.[28] At the same time, because the Allied experts had accepted the fundamental premise of the German position that the major successors to the I.G. would have to be large enough to compete effectively on world markets, negotiations between the Allies and the Germans could proceed by the end of 1951 and beginning of 1952 to concrete results. The major successors to the I.G. (Hoechst, BASF, and Bayer) as well as a number of smaller successors, officially came into existence in this period. One might even say that the three major I.G. successors were "refounded" since each existed prior to the founding of the I.G. and continued to exist as a works group in the I.G. era. Most other issues were settled by the time the law for the conclusion of the I.G. liquidation took effect on February 6, 1955, although court battles dealing with aspects of the disentanglement (e.g., properties in the "Soviet zone," suits brought by slave laborers, and so forth) continue to be fought by the legal successor to the I.G.—I.G. Farbenindustrie A.G. in Abwicklung in Frankfurt.

The most important treatment of this subject is Hans-Dieter Kreikamp's article on "The Disentanglement of I.G. Farbenindustrie A.G. and the Founding of the Successor Companies."[29] Kreikamp stresses several major themes and arguments in his analysis, and one must agree with his conclusions for the most part. For one thing, he emphasizes the influence of American military government personnel—and especially those advocating the ideas of decartelization and trust-busting—on the disentanglement process.[30] Second, he observes with justification that the factor of time helps explain why the breakup was not more radical than it was. Had the Allies acted on their plans immediately after the beginning of the occupation (or within the first two or three years), they may have been able to deconcentrate the corporation more completely into smaller units.[31] Third, absence of interest-group conflict on the German side was an important factor in explaining their influence on the proceedings. Federal bureaucrats, industrialists, and trade unionists were united in their desire to thwart radical deconcentration of the I.G.[32] Finally, Kreikamp points out that technical arguments used in the negotiations were fundamentally ambiguous. These arguments, though couched in technical language (e.g., interdependence of production among various factories, optimal firm size, and so on), were not fully convincing since both sides could marshall evidence to support their respective interests. Instead, the arguments were at their core disagreements over basic economic policy objectives.[33] The outcome of the negotiations was to a large extent a return to prewar property and economic relationships.

There are two major shortcomings in Kreikamp's otherwise excellent analysis. First, he tends to underestimate the extent (and therefore the effects) of ambivalence within the ranks of U.S. military government personnel. As outlined in chapter 2, there were two main groups within the military government. On the one hand, men from the Antitrust Division of the U.S. Department of Justice and from the Treasury Department dominated the staffs of the decartelization division and the I.G. Control Office. They tended to favor radical deconcentration of the German economy. On the other hand, many of the other staff members of the economics and finance divisions of the military government were representatives of big business interests. At the beginning of the occupation, men from the first group influenced U.S. policy most heavily. As the occupation period went on, those from the second group gained the upper hand in general. Still, it is telling that a new staff representing the interests of the antitrust faction was appointed to man the decartelization commission of the U.S. High Commission of Germany at the beginning of 1950.[34] In effect, U.S. military government in Germany represented *throughout the occupation period* (1945–1955) the ambivalence toward big business that has characterized *American* economic history.[35] The compromise position on the I.G. eventually accepted by the American authorities in Germany— that is, limited deconcentration and preservation of competition through establishment of an oligopoly in the chemical industry— also mirrored the outcome of such struggles within U.S. government.[36]

The second shortcoming of Kreikamp's analysis is his periodization of the disentanglement process, and in particular his identification of its "critical period." This study has demonstrated that the very fact that Germany was broken up into four airtight zones began the process of disentanglement. What is more, French and British policy tended to favor centralization, at least at the level of the works group. Only the Americans went further in their disentanglement of production relationships within the former I.G. Retention of the works groups in two of the three western zones was bound to weigh heavily in any discussion of the formal and legal breakup of the firm. Kreikamp himself notes that even at the beginning of the "critical period" in the disentanglement (summer and fall 1950), because of the size of the two groups, BASF and Bayer, "both sides [Allied and German] tried to limit the separations from the Maingau complex and instead to attach several individual companies to the central works of Farbwerke Hoechst."[37] One might thus claim that the "critical period" for the determination of the overall shape of the disentanglement was during 1945 instead of 1950, and in any case long before the details of the disentanglement were formally resolved.

There is yet another problem in dating the disentanglement in 1950–1951, and consequently in seeing these years as crucial ones in the reconstruction of the former I.G. firms. After all, each of the three major factory groups of the former I.G. began the process of reconstruction long before this date. Production levels, by 1950, approached—or even surpassed—those of the prewar years, and the successor firms had begun to export at a rate approaching the boom years of the 1920s (in percentage of turnover derived from export). Formal disentanglement in the period 1951/1952–1955 was thus not a key factor in the recovery of the industry. One must agree with Kreikamp, however, that the formal breakup in the early 1950s was crucial for the resurgence of the former I.G. firms. The founding of the successor firms in 1951 and 1952 and the subsequent settlement by 1955 of most outstanding issues surrounding the disentanglement meant elimination of an added element of confusion and uncertainty for the companies. They were thus able to turn to directing and managing the resurgence of the West German chemical industry.

THE BASES OF RESURGENCE: THE PERSONNEL AND PROSPECTS OF THE "NEW COMPANIES"

Even though the final disposition of the I.G. was not settled until the mid-1950s, all parties had moved close enough by December 1951/January 1952 to permit provisional founding of the three major successor companies. The first of the successors, Farbwerke Hoechst Aktiengesellschaft vormals Meister Lucius & Brüning, came into existence on December 7, 1951. Consisting of major factories at Hoechst, Griesheim, and Offenbach, with key subsidiaries in Knapsack and at the Behringwerke in Marburg, the new company resembled in its main contours the former Maingau production unit of the old I.G.[38]

The other major successors came into existence shortly thereafter. Farbenfabrik Bayer A.G. was founded on December 19, 1951, consisting of factories at Leverkusen, Elberfeld, Uerdingen, and Dormagen.[39] Later the Agfa complexes at Leverkusen and Munich were added to the firm as 100-percent subsidiaries. The new Bayer firm reproduced almost precisely the former Lower Rhine Group of the I.G. On January 30, 1952, the final major successor came into being. Named BASF, the new firm consisted primarily of the works at Ludwigshafen and Oppau, with some important stock participations in other former I.G. firms.[40] BASF was in essence the I.G. Upper Rhine Group without the factories in Poland or in the "Soviet zone of occupation" (East Germany).

Each of the new firms thus conformed to a remarkable degree to the organization of production during the I.G. period, and even prior to that when each had existed as an independent company. They shared other similarities. One of the most striking was that each of the new firms came into existence with a basic capital of DM 100,000. Obviously the nominal stock value of the companies had nothing to do with their actual value, but rather reflected the political decision on the part of all participants to complete the process of breaking up the I.G. as quickly as possible. It meant, moreover, that all participants agreed on basic issues surrounding the question of the structure of the West German chemical industry in the postwar world. Allied and German representatives accepted the figure of DM 100,000 since it was necessary to create a legal entity (*Rechtspersönlichkeit*) to take over the property and functions of each of the major successors. In order to comply with German law, at least five men had to form each of the companies. The men had, furthermore, to come up with the nominal founding capital on their own. DM 100,000 thus was a convenient and affordable amount for each of the groups of five to found the new companies.[41]

Creation of the new firms as legal entities allowed final negotiations to take place over the capitalization of the three leading chemical firms in West Germany. Karl Winnacker, director of the Hoechst factory from 1943 until his removal by the Americans in 1945, and named chairman of the board (*Vorstandsvorsitzende*) by the supervisory board of the new firm on April 2, 1952, described the negotiations well in his memoirs. In December 1952, he wrote, representatives of the three major successor organizations, two representatives of the federal government (Prentzel and von Rospatt), several members of the stockholders' advisory council, and the liquidators of the I.G. (in the I.G. Farben Liquidation Committee) met to divide up the capital of the old I.G. The discussions reflected the real conflict between the interests of the successor firms themselves and those of former I.G. shareholders, who would of course receive shares in the successor firms. Shareholders, Winnacker noted, had an interest in relatively high capitalization, whereas each of the firms had an interest in low capitalization. There were good reasons for this:

> The magnitude of the [I.G.] share capital taken over by [the individual firms] (or perhaps better put, imposed upon them) determined of course the extent of their obligations to the later shareholders. Thus he who received less than his competitor in the course of the division possessed later better possibilities for finding financing.[42]

Each major successor ruthlessly pursued its own interests in the negotiations that followed. Winnacker proudly walked away from the negotiations with relatively low capitalization for Farbwerke Hoechst. Most of the capital of the I.G. was divided as follows: Hoechst, DM 285.7 million; BASF, DM 340 million; Farbenfabriken Bayer, DM 387.7 million; and Casella, DM 34.1 million.[43] The three successors were "refounded" (the Germans called it "*Ausgründung*" or "*Nachgründung*")[44] at the end of March 1953.

The three faced the prospect of continuing reconstruction and recapturing world market share differently after the official breakup of the old chemical giant. Bayer, for instance, emerged from the disentanglement with a major asset, the "Bayer cross" trademark. There was probably no more well-known trade name in the chemical industry. The I.G. had, moreover, marketed all of its extremely profitable pharmaceuticals—including, for instance, those manufactured in the Maingau Group's Behringwerke plant in Marburg—which made the possession of the trademark an invaluable asset in regaining former markets. Extensive technical integration among Bayer's four main plants was a further advantage. Finally, Bayer had strong management, largely intact physical plant, an active research effort, and a strong tradition of salesmanship. A study published in mid-1952 by the Spezial-Archiv der deutschen Wirtschaft and distributed by the successor of the Dresdner Bank, the Rhein-Ruhr Bank, was positive in its estimation of Bayer's future chances:

> In general, the future development of turnover for the Farbenfabrik Bayer is . . . judged to be stable.
> It is of particular significance for the future development of the Far-

benfabrik Bayer that the reconstruction of the research facilities is quite far along, and that research is being pursued with every intensity in areas of interest to the company. Since the currency reform, a number of valuable new products have appeared on the market as a result of its own research.[45]

BASF in Ludwigshafen shared with Bayer the advantage of technical integration between its two main facilities in Ludwigshafen and Oppau. It shared, furthermore, other Bayer characteristics, including strong management and an active and successful research tradition. On the other hand, the two main plants of the Upper Rhine Group had been among the most heavily damaged I.G. factories during the war, and the catastrophic explosion of July 1948 not only undid some rebuilding efforts to that point, but also placed the Ludwigshafen facility at a lower capacity than it had been since the end of the conflict. The Spezial-Archiv was thus more cautious in its predictions for the prospects of the "new" BASF:

> If the upward trajectory of sales in the first half of 1951 (which resulted from special circumstances) has since experienced a certain slowdown, prospects for further development of production and markets are still not unfavorable. In the coming years, removal of remaining damages due to war, dismantling, and the explosion will still naturally require considerable effort and financial means. The closer reconstruction gets to its conclusion, the stronger the expectation that long-range planning in connection with the results of research activities of the company will have a positive effect on turnover and result [sic].[46]

Factories of the Farbwerke Hoechst A.G. had suffered very little damage during the war. Nonetheless, considerable pessimism surrounded the future prospects of the company: physical plant tended to be fairly old; it was the least integrated of the successor companies; and management problems had plagued the firm (in the Maingau, or Middle Rhine Group) for some time. Spezial-Archiv der deutschen Wirtschaft analysts were cautiously optimistic about the firm's chances, but placed most of their hopes on the future:

> The further development of production and the winning back of lost ground on the world market can only be built upon the foundation of extensive research of its own which includes new areas and further

develops old ones. In recognition of this, special emphasis was placed in all factories on expansion of the numerous laboratories, experimental facilities, and applied technical sections, which have been enlarged and the equipment of which has been steadily supplemented and expanded. A large number of scientists and technicians, together with an enlarged staff, has been brought to bear on research and development work. The R & D work concerns all branches of the [firm's] extensive production and its practical application.[47]

Each of the three major successors to the I.G. possessed however an important set of assets that boded well for its future business prospects: personnel. For one thing, each of the chairmen of the managing boards had been in a similar position *in the same production group* during the I.G. period. Ulrich Haberland became director of the Lower Rhine Group in 1943, while Carl Wurster was already director of the Upper Rhine Group in the late 1930s. Karl Winnacker of Hoechst had been a prominent manager in the largest of the facilities of the Maingau Group beginning in 1943.

What is more, it appears that all of the members of the managing boards had been, in general, employees of the particular I.G. production group for which they now became directors as well. It is unclear how far down the management hierarchy this characteristic holds true. For upper management it is certain, however. At least five of the eight members of BASF's managing board in the immediate post-breakup period, for instance, were employees of the Upper Rhine Group during the war. These included Wurster and his deputy, Bernhard Timm (later chairman of the managing board). Timm, originally trained as a physicist, had begun his career as an astronomer in the personal employ of Carl Bosch, one of the most important figures in the history of BASF. Bosch brought him into the BASF works and promoted his career within it.[48] Wolfgang Heintzeler, a lawyer, was also with the group during the war. He had shown his loyalty to the director of the factory complex by participating in the defense of Wurster at Nuremberg.[49] Dr.-Ing. Walter Ludewig and Prof. Dr. Walter Reppe were both prominent in the technical staff of the Upper Rhine Group before and during the war.[50] At least twelve of fifteen members of the managing board of Bayer in the early 1950s had been in the Lower Rhine Group of the I.G., and at least four of eight members of the Hoechst managing board were managers in the Maingau Group.

The implications were of great importance for the recovery and resurgence of the major I.G. successor firms. Each of the three possessed an upper level of management familiar with the particular problems and potentialities of the given production group. In the case of BASF and of Bayer, what is more, these men were the products of a conspicuously successful system of production units. The Maingau Group, which later became for all practical purposes Hoechst A.G., had a less impressive record of achievement within the I.G. The new chairman of the managing board of the company was, however, different from the production group managers who had preceded him. During his tenure as chairman during the 1950s, Karl Winnacker promoted remarkably young (and capable) people within Hoechst to positions of influence. Kurt Lanz, a member of the managing board later in the 1950s and instrumental in reconstructing Hoechst's world trade, was in his early thirties when promoted to the upper management, and just 39 when first appointed to the managing board of the company. The man who later replaced Winnacker as chairman of the Hoechst managing board, Rolf Sammet, was a similar case.[51]

Return to world competitiveness on the part of new companies was not simply the result of a shared characteristic of high-caliber personnel. It was also fortuitous that the three chairman of the board and their fellow board members presided over their firms at a time when the world economy had begun the largest and longest period of economic expansion the world had ever seen.[52] In the meantime, they all had to deal with the problems of the 1950s for the chemical industry in West Germany, for not all was rosy during these years. One of the most important problems they faced, and overcame, was that of finding the money necessary to rebuild and expand. The fact that they could find the necessary capital was also a factor in explaining the resurgence of the industry during the 1950s.

THE BASES OF RESURGENCE: INVESTMENT AND CAPITAL DURING THE 1950s

One of the most important questions one must confront in trying to assess the reasons for the resurgence of the West German chemical industry, and especially of the I.G. successors, is: where did the money come from? Without the capital necessary to rebuild plants, to design and purchase more up-to-date equipment, to set up sales representation abroad, and to conduct necessary research on new products, all of the other qualities of the I.G. successors—a strong tradition of excellence, high-caliber management, and large-scale production capacity—were of limited use. The answer to the question is a difficult and complex one.

By the end of the war, the factories of the I.G. were only partially destroyed. Nonetheless, destruction was considerable, and dismantling and reparations took away still more physical capacity from the German chemical industry. In addition, one-half to two-thirds of the company's holdings disappeared through loss of territories in middle and eastern Germany and through forfeiture of foreign holdings. Remaining factories were often worn out and technologically backward. The three western successors were nonetheless able to rebuild with astonishing rapidity. Indeed, in 1953, the three large successors to the I.G. were able to post a combined turnover of DM 2.853 billion, over 90 percent of the 1943 corporate high turnover of RM 3.116 billion. The fact that the later figures are in deutsche marks (rather than in inflated reichsmarks) makes the recovery still more impressive.[53] How was it possible?

The chemical industries section of the Federal Ministry of Economics collected statistics on the industry's investment needs. As of March 1953, it had registered requirements of DM 541 million since the beginning of aid under the European Recovery Program (1948). German-state and American programs contributed approximately DM 198 million toward fulfilling these needs. Other public programs would contribute about DM 43 million more. Thus West German and American investment programs satisfied about DM 241 million of the investment capital requested by the West German chemical industry. It is not clear which part of the West German/American total would come from European Recovery Program (as counterpart funds) and other U.S. aid programs (for the most part STEG, a program to make surplus army stores available to the German economy that also had a capital investment component). But the programs that were indisputably American would contribute a total of DM 120.96 million toward investment needs in the industry, in other words more than 50 percent of the West German/American total.[54]

Aid to the West German chemical industry given by the government of the new Federal Republic and by the Americans was no doubt crucial to explaining the fact that sufficient funding existed for its resurgence. But capital needs of DM 300 million—more than half of the industry's total capital demand—were left unfulfilled from public sources. Where did the remainder of the funds come from? Banks, of course, were instrumental in providing part of the remaining investment needs. But, as we have seen, funds from the banks were insufficient as well.

Remaining investment capital for the chemical industry came from within the industry itself. But the industry was so obviously troubled with liquidity problems in this period that it is not immediately evident how it could afford such massive internal financing in order to rebuild and to modernize. One key source of funds has been described well in a little-known book by Karl-Heinz Forster, published in 1953 as *Finanzierung durch Abschreibungen: Nach den Ergebnissen von DM-Bilanzen*.[55] Forster argued that prior to the currency reform in mid-1948, the major problem for industry was finding raw materials for production. With the enactment of the reform, the primary problem became instead finding sufficient capital for investment in reconstruction. But capital was scarce:

> In view of the impoverishment [through the currency reform] of the social strata which had savings and the lack of those capital funds as well as the absence of long-term foreign capital, it remained the task of industry to implement financing on its own.[56]

The years following the currency reform saw a resurgence in profits for industry in general. On the other hand, there were also high rates of taxation. The result, Forster argued convincingly, was that tax write-offs (*Abschreibungen*) were of immense importance. Forster claimed that, "through write-offs, capital is turned liquid and temporarily—or even permanently—freed up and thus can be applied for other tasks of financing. . . ." He concluded: "We must therefore speak of write-offs as a source of financing."[57]

But where did the write-offs come from? Forster argued that the Deutsche Mark Balance Law (DM-Bilanzgesetz), which took effect on August 22, 1949,[58] played a key role here. It had a vey important result:

namely, the partial renewal, increase, and modernization of industrial potential through the release of capital that was bound up in the plant available on 20.VI.1948, but not expressed in accounting terms. What this means can only be measured if one takes into consideration the condition of our industrial plant in the middle of 1948, the low production index in the initial postwar years, the limitless capital poverty shortly after the currency reform, and the high tax burden.

The write-off of industrial plant therefore—and that is the most important result of our investigation—in large part made possible the reconstruction of the physical plant of our economy, which had suffered serious destruction during the war and the ensuring years. The upswing of the German economy, in particular the speed with which it took place, is thus due not least of all to the DM-Bilanzgesetz.[59]

Forster contended that wealth available to German industry (because its facilities were to some extent intact after the war) was released in part through legal means. Facilities with previously high reichsmark values could now be valued in deutsche marks because of the law. In other words, an accounting trick freed vast amounts of capital of immense value to the German recovery. Since the facilities of the German chemical industry, and in particular the factories of the I.G., were surprisingly intact at war's end, and since the I.G. successors lost significant holdings in eastern Germany, Poland, and overseas, they were prime candidates for the use of tax write-offs.

The final factor accounting for the availability of sufficient capital to finance the rapid expansion of German industry applied primarily to the I.G. successors and is more difficult to demonstrate. It is nonetheless probable that systematic undercapitalization of the firms from the days of the I.G. allowed them to be able to find investors to expand the capital bases of the successor firms, and thus to finance investment projects. To support this contention, it is necessary to digress briefly.

In order to understand the "undercapitalization" of the I.G. and its successors, it is important to keep in mind a very simple fact: capital shares of, say, RM 1000 represented a *debt* for the firm of RM 1000.[60] Thus capital is to some extent a two-edged sword. It is necessary for the formation of a very large firm, but too much of it means that the firm has to pay enormous "interest" on its "debt":

the firm had to pay out dividends to stockholders and to follow their wishes to some degree if heavily in "debt." The concepts of over- and undercapitalization follow from this. An undercapitalized firm does not have access in general to the amount of investment capital it could command on the stock market. On the other hand, that same firm has more control over its own affairs (i.e., the managers have more control than the stockholders), and it need not pay out so much of its profits in dividends. The overcapitalized firm has access to a very large amount of capital for investment in plant expansion and acquisitions. But it must pay for these advantages dearly through loss of some management control, and through paying out profits as dividends.

It appears that the I.G. was undercapitalized. And such a strategy is possible, perhaps even wise, for a firm with large profits, which the I.G. was. The book value of the firm at the fusion in 1925 was RM 646 million, climbing by 1926 to RM 1.1 billion. The market value of the stock was over RM 3 billion in 1928.[61] Now this was a very large corporation indeed. Helge Pross, for instance, estimated that the average corporation in Germany at the end of 1925 (including banks apparently) had a nominal capital of RM 1.5 *million*.[62] What is more, the I.G. paid quite well for the use of this capital, since dividends amounted in 1926 to 10 percent of nominal share value, to 12 percent from 1927 to 1930 (plus a bonus of 2 percent in 1929), to 7 percent from 1931 to 1936, and to 8 percent in 1938.[63]

But the company's stock policy as it developed in the years after fusion shows that the thesis of undercapitalization is justified. First of all, the I.G. owned a substantial portion of its own stock. A certain amount was issued in the late 1920s to help pay for the construction of the new administration building in Frankfurt. But the firm bought back many of the shares in 1930–1931 when their market price fell. In late 1951/early 1952, the nominal value of the stock was fixed at RM 1.36 billion by I.G. liquidators.[64] This means that the firm's nominal capital was only slightly higher at the end of its existence than it was in 1926. Experts nonetheless estimated the value of the prewar holdings and property (*Vorkriegsvermögen*) at RM 6 to 7 billion.[65]

The I.G. lost considerable property and holdings in the eastern zone and abroad after the war. There was also substantial bombing damage. Nevertheless, the value of of the firm's assets *in the western zones alone* was more than the stock value of the concern:

> The net worth of I.G. Farbenindustrie amounted to DM 1,479,228,525.40 on 21 June 1948, the key date of the currency reform, with a nominal capital of RM 1,360,000,000. in common stock [*Stammaktien*] (40 million preferred stocks are in the possession of I.G. Farbenindustrie). Despite the extraordinarily high war losses, . . . net worth thus . . . still surpasses somewhat the original nominal capital.[66]

In sum, it may be posited with great probability that the firms in the western zones after the war were *still* undercapitalized, given their property, personnel, and know-how, and Hoechst may well have been in the best position of the three major successors from this standpoint. The factor of undercapitalization was crucial to explaining the resurgence of the firms. After the breakup of the I.G. and the settling of its "debt" to stockholders, all three major successors were in the position of going to the capital markets in the firm conviction that they had something to offer new shareholders.

And they did go to the stock market to try to attract new shareholders, with some success. BASF raised its basic capital (*Grundkapital*) by nearly 50 percent in 1955, to DM 510 million. Farbenfabriken Bayer's management increased their nominal capital to DM 550 million in the same year, an increase of about 42 percent over the original worth of the firm at the refounding in March 1953. Hoechst did the same thing, increasing its capital to 385 million in early 1955, an increase of about 35 percent. Almost all of the increase in capital in each case went toward financing expansion of existing facilities and long-term investment in new facilities and research.[67]

THE BASES OF RESURGENCE: PARTICIPATION OF THE I.G. SUCCESSORS IN THE U.S. POSTWAR ECONOMIC SYSTEM

There is a final major factor in explaining the rapid recovery and expansion of the West German chemical industry in the 1950s, and that is its participation in the postwar U.S. economic order. To some extent this was due to the remarkably parallel interests of the German firms on the one hand and of U.S. policymakers on the other: German chemical industrialists had long recognized that recovery and improvement of the industry's prospects lay in foreign markets and that, without strong exports, the industry could not expect to survive in the postwar world; many U.S. policymakers were interested precisely in the same thing. They believed that the postwar world would have to be based on strong and free trade among all nations if another depression were to be avoided. What is more, a Germany dependent on foreign trade rather than on attempting to supply its own needs through a policy of autarky would of necessity be more interested in maintaining the peace than in disrupting it. Finally, the developing Cold War led U.S. and other Allied policymakers to favor an economically and politically strong western Germany, which would by implication be heavily reliant upon export.[68]

The process by which the new West German state began to participate on a basis of equality with other states in the world economic system took some time and was composed of two important sets of developments. The first was the reestablishment of the legal framework for the conduct of trade through the renewal of trade and business protection measures (e.g., trade marks and patents) in West Germany. The second was the renewal of the institutional framework in which industry could conduct trade. For the chemical industry, the most important elements in this process were the establishment of representation abroad, the beginnings of European cooperation, and the establishment—and sometimes reestablishment—of contacts with U.S. firms.

Initially, the Allies seized German industry's patents and prohibited use of its trademarks until agreement came on their ultimate disposition. Trademarks were a very important sales tool in certain segments of the chemical industry, including, above all, pharmaceuticals. At first, Allied policy had the effect of disallowing their use by German firms. For instance, U.S. military government personnel, in a directive to the Hessian economics ministry on the proposed participation of former I.G. plants in a trade show in Wiesbaden, ordered that,

> In general, trade names and trade marks affiliated with the former I.G. Farben industry should be excluded. The products on display should be offered, wherever at all possible, as emanating from an independent plant. . . .[69]

British policy was similar, although based less on the desire to limit concentrations of economic power on the part of the I.G. than on the intention to avoid giving German industry advantages over British industry. In a minute discussing the proposed policy of either wiping out German trade marks altogether, or of allowing all who wish to use them, A. L. Burgess of the Board of Trade offered these observations:

> I personally think that the proposed policy is right, and not in any way inconsistent with our desire that the Germans shall be able to export. The fact is that, in fields where trade marks really matter, e.g. dyes, drugs and chemicals, whoever gets in first and gets a trade mark of the "Luminal" [a Bayer product] type established in the public and trade mind, tends to get a world monopoly of the trade.... [I]n the organic chemical industry it was the Germans who got in first, though British chemists made the fundamental discoveries. I agree that the Germans must be allowed to sell chemicals and pharmaceuticals in competition with us, but I see no reason why we should hand back to them the trade marks which would largely prevent us from competing with them.[70]

Thus, in developing a "Plan for the Dissolution of Bayer Sales Organization" in March 1946, the OMGUS Decartelization Branch devoted some care to discussing the disposition of trade marks and patents. The representatives of the Branch suggested that the Bayer Sales Organization, which was the exclusive marketer of I.G. Farben pharmaceuticals, should be broken up into its component parts and sold. The language in the plan was terse:

> Sell the name Bayer and the Bayer Cross with the sales company.
> Sell the name Emil von Behring with the Behringwerke, Marburg.
> Sell the product trade names with the plant producing the finished product.[71]

The patents belonging to the sales organization and its component factories, the report continued, "will be treated in the same manner as other German patents since pharmaceutical patents enjoy no particular distinction among chemical patents as a whole."[72]

The treatment of "other German patents" referred to in the report was harsh. Diplomats from all over the world participated in a successful effort to formulate an accord on German patents. All German patents seized by August 1, 1946, were to be available either to the public at large or without royalty to the nationals of the signatory countries. The patent accord furthermore set up a central office in France to administer and coordinate the implementation of the agreement, and to ensure protection of the property of German refugees and non-Germans. Representatives from the signatory nations gathered to initial the treaty in London on July 27, 1946, and it came into force on November 30.[73] In the meantime Germans could not, and would not in the near future, be able to register any products or processes discovered after August 1, 1946, since the old Berlin patent office had been closed down. Temporary "registry offices" came into existence at the *Land* level in the immediate postwar period. But the Americans and British closed these down in early 1946 out of the desire to treat Germany as a whole in this matter. Until 1948, there was no legal way to establish patent priority.[74]

The issues of patent and trade-mark protection were thus bound up with other major issues in the occupation: Should Germany be treated as a whole economically and politically? How was the breakup of the concentration of economic power in such companies as I.G. Farben to proceed? To what extend should Germany industry be crippled in order to undermine potential military or commercial threats from the defeated nation? Resolution of the issues of patent and trade-mark protection came about in conjunction with the resolution of other major issues. The question of trade marks for the I.G., for instance, was resolved in the course of the disentanglement of the company. In general, the firm given ultimate possession of valuable I.G. trade marks was the same as the one given ultimate possession of the factory producing the good in question. When many factories produced a particular patented good, the solution varied. Bayer, for instance, received the exclusive right to utilize the world-renowned Bayer cross.[75] On the other hand, as part of the negotiations on the breakup, all three successors to the I.G. (as well as other companies including Francolor, S.A.) licensed prior to or during the war to produce "Indanthren" dyestuffs, received the right to use that valuable trade mark. They set up the Indanthren-Warenzeichenverband e.V. in 1952 to administer it.[76]

The beginnings of normalization of the patent situation in western Germany came earlier. Already in April 1946 the French began to allow registration of German patents in their country. The United States and Great Britain followed with laws allowing for the registry of patents for goods or processes discovered after the beginning of 1946 on August 6, 1947, and April 8, 1948, respectively. But the benefits of this legislation were questionable. It was not until September 9, 1948, for instance, that the bizonal Joint Export–Import Agency with JEIA Directive No. 24 allowed implementation of the agreements from the German side. Even then, the export control system did not support financially registry of patents abroad; there was also no provision in the directive for the coverage of costs incurred through patent litigation abroad.[77]

Establishment of a patent office in Germany itself remained an unresolved issue, particularly since it was tied to the issue of reunification. Geman policymakers discussing a proposed law to establish an office to register patents, trade marks, and other trade protection measures in the spring of 1948 argued for the necessity of patent and trade-mark protection above the *Land* level. At the same time, they argued that such protection should be given through a central patent office for all of Germany. Thus a Patent Office for the bizone or for the western zones should not be erected until the possibility of creating a Central Office was closed out entirely. Until then, registry offices (*Annahmestellen*) would be set up only for the bizonal area.[78] Two offices, one in Berlin and one in Frankfurt, thus began functioning in October 1948.[79]

The German Patent Office itself did not come into existence until October 1949, when it was established in Munich and the two registry offices closed down. As the date implies, establishment of this office had to do with the establishment of a West German state. A conversation in April 1949 between Hermann Pünder, the executive director of the German bizonal administration, and U.S. Military Governor General Lucius Clay, is indicative of the complexity of the problem and its relationship to the establishment of the German state. Pünder asked Clay about the prospects for permission to erect a Patent Office. Clay replied that, unfortunately, no news on the problem was available because of discussions going on at the highest levels. He added that, "The simplest way [to resolve this issue] would be for you to erect a German regime soon." Pünder thereupon asked, "May we then now proceed with the preparations for the establishment of the Patent Office? To know that is for us of great significance." Clay responded, "Yes, if you don't consult us about it."[80]

As the major political problems of the occupation were solved, so were the issues related to trademark and patent protection. By the early 1950s German business was on equal footing with other nations both with regard to protection at home and abroad. The resolution of these issues and the reestablishment of trade protection were of immense importance to the I.G. successors in their effort to rebuild and expand during the 1950s.

The second set of developments through which firms in the West German chemical industry began to participate in world trade during the 1950s had to do with the renewal of the institutional framework necessary to trade development. Establishment of representation for the firms abroad, growing European cooperation in the chemical industry, and attainment of agreements with U.S. firms in the chemical and petroleum industries all were important here.

In the immediate postwar years, trade had been suspended for all practical purposes for the western German chemical industry. Production levels were low, demand at home was high, and in any case the Allies imposed restrictions on chemical trade and production. For the I.G., all foreign subsidiaries, including sales companies abroad, had been seized during or after the war. Finally, there was some backlash in the world market to German goods following World War II, and this backlash applied in particular to a firm such as the I.G., which was so heavily implicated in the Nazi regime and its crimes.

It is clear, however, that considerable demand remained for German chemicals, and in particular those produced by the I.G.[81] What is more, Farben's successors recognized acutely their dependence on foreign markets and moved as soon as possible to reestablish, or to establish representations in lucrative foreign markets. By the early 1950s, Allied restrictions on this were relaxed or abolished. The remaining constraint on German efforts was insufficient foreign exchange to establish sales organizations abroad. Nonetheless, I.G. successors made headway relatively rapidly in the area of foreign trade, as the stories of two major figures in the industry illustrated.

New Beginnings and Resurgence 195

Kurt Lanz was one of the talented young men who would ensure that Hoechst would compete with the other two major successors to the I.G. in Germany and abroad. Born in Mannheim in 1919, he went to work for the I.G. in Frankfurt in 1937. Lanz was not trained as a chemist, but rather was attractive to the export-oriented I.G. because of his abilities in languages. After a stint in the main factory of the Maingau Group, Lanz was called to work in the language service of the *Wehrmacht* from 1940 to 1945. Because of his facility in languages, previous connection to the I.G., and lack of affiliation with the Nazi party, Lanz had no trouble securing a position in the Griesheim works of the former I.G. in September 1945. He moved to the economics section of the Hoechst works in May 1946. It was here, during the years that followed, that Lanz and others slowly and laboriously built up an extensive foreign sales operation for the Hoechst works and then for Hoechst A.G. Lanz and other Hoechst sales representatives utilized former associates of the I.G. in foreign countries, sales and technical agreements with foreign firms, and—as soon as possible—extensive foreign travel to set up foreign sales outlets for Hoechst. By the early 1960s, the efforts paid off handsomely with representations the world over for the company.[82]

The Bayer organization had numerous men who aggressively tackled the problem of setting up foreign sales organizations as well. One of them, Kurt Hansen, later became the chairman of the managing board of Bayer A.G. Hansen himself was a product of the German "commercial tradition." Born in Japan in 1910, the son of a German sales representative and his wife (who herself came from a long-standing family of salesmen), Hansen traveled with his mother in 1913 across Russia to Germany on the Trans-Siberian railway. Hansen was already a Diplom-Chemiker at the age of 23, and, in 1935, took his doctorate in Munich. He was hired by I.G. Farbenindustrie A.G. in the following year, worked as a research chemist in Leverkusen, spent some time in the *Wehrmacht*, and then became a procurement representative for the I.G. in Berlin.

Hansen returned to Bayer after the war and took up work as a research chemist for the factories of the old Lower Rhine Group. In the early 1950s, however, he was chosen to set up the important sales representation for the Bayer corporation in New York. The firm lacked dollars to do an adequate job at first, and Hansen was thus forced to become a one-man sales operation for the company in a small and poorly equipped New York office.[83] Demand for chemical products of all sorts—in part as a result of the Korea boom—and demand in particular for the highly regarded products of the former I.G. firms eased Hansen's (and Lanz's) task, however. It was not long before export to lucrative markets such as America was a vital part of the business of the successors. By 1955, because of the efforts of these and other men, the successors to Farben had sales representations throughout the world and important subsidiaries, especially in the United States and South America.[84]

But resumption of activity on the part of the successor companies in the area of overseas representation, while important, was not the sole component of the institutional reconstruction of the firms. Another vital component was cooperation with European firms. The most important market for German industrial products, including chemicals, after all, was and would continue to be the neighbors of West Germany in western Europe. And the most important of these neighbors—in economic, political, diplomatic terms—was France. German chemical industrialists took up contact with their French counterparts at the first real opportunity after the formation of the Federal Republic, but before the final breakup of the I.G. The German newspaper *Die Welt* reported on a "Franco-German chemical discussion" on June 7, 1950:

> Confidential discussions between representatives of French chemical trade association and the *Arbeitsgemeinschaft Chemische Industrie* [the Chemical Industry Trade Association] took place recently in Frankfurt. The topic of the discussions was the reestablishment of good prewar relations between the chemical industries of the two countries. Especially highlighted was the necessity of comprehensive exchange of information. What is more, in the future, fundamental problems of European and international trade insofar as they affect the chemical industry are supposed to be discussed in common.
>
> As reported, the first contacts proceeded satisfactorily. The occasion for the French visits seems to have been objective of checking out the possibilities for a Franco-German combination of the chemical industry on the model of the planned coal and steel union.[85]

A meeting of West German chemical industrialists later in the summer demonstrated that they expected a great deal from the much-touted European cooperation, and noted that the aforementioned conversations with the French "had shown no fundamental differences." They were, however, cautious about the idea of European cooperation since "it is clear that the fulfillment of European tasks by the west German chemical economy just now involves more victims than advantaged."[86]

The idea of a European Chemical Community alongside the European Coal and Steel Community is an arresting one. Why, in fact, did the latter come into existence while the former did not? The situations were, after all, similar in one important way: a strong German chemical industry faced a rather weak French counterpart, just as the German coal and steel industry was far stronger than French heavy industry. There were several reasons why a European Chemical Community did not come about. For one thing, chemical production was by no means as important either symbolically or—at least in the short term—economically as the coal and steel industry. Then again, French chemical interests were able to gain favorable contracts with German producers as a result of the occupation and the disentanglement.[87] The French, after all, administered a significant part of the chemical industry— but not of the coal or steel industry—in their zone.

I would like to suggest, however, that there is one further important reason for the fact that there was no equivalent to the ECSC for the chemical industry: agreements with U.S. chemical and petroleum firms. I.G. successors began to sign pacts to share technology with American firms even before the breakup was final. Hoechst, for instance, engaged in a major contract with Merck & Co. in Rahway, New Jersey, to produce large amounts of penicillin via the Merck process at the behest of the U.S. military government.[88] Farbenfabriken Bayer signed a more equal technology sharing/sales agreement with Schenley Laboratories, Inc., a major pharmaceuticals producer. The two firms agreed to sell each other's products in their respective countries, and to embark on a cooperative research program.[89]

Beginning with a contract signed in 1953 between Shell Oil and BASF on a joint petrochemical venture in West Germany, I.G. successors moved rapidly into the petrochemical technology needed to compete on the world market.[90] By signing contracts with large oil companies, the firms were able to avail themselves of the latest technologies, but the contracts also had the effect of making the German chemical industry even more dependent on world markets: the industry would have to look abroad now not only for its customers but also for its supplies of raw materials. This development—the major change in the technology and raw materials base for the German chemical industry during the 1950s—helps explain why there was no chemical parallel to the ECSC. Since West German chemical industrialists were dependent upon U.S. and British oil companies for supplies, there was no need to control the industry through some sort of supranational authority.[91]

Clearly, a number of factors contributed to the "take-off to sustained growth" of the production of the West German chemical industry, and in particular of the I.G. successors, during the early 1950s. Much of the expansion was based on the explosion of world trade following the war, especially in the aftermath of the Korean War. The political–power constellation, West Germany's growing sovereignty, and the country's place in the U.S. postwar economic order were thus all factors in the industry's resurgence. Farben's successors benefited, in addition, from resolution of the breakup of the old firm, from their personnel and traditions carried over from the I.G. (and to some degree even the pre-I.G.) period, and from availability of sufficient investment capital with which to prepare the industry for resurgence.

PART IV
SUMMARY AND CONCLUSIONS

With the founding of the major successors to I.G. Farbenindustrie A.G. in late 1951 and early 1952, a new era dawned for large-scale German chemical production. Instead of a single firm that dominated the most lucrative, technologically advanced, and export-oriented segments of the industry, three corporations of approximately equal size and capability competed vigorously with one another in domestic and foreign markets. The following chapter surveys how this change came about and its implications for West German business and politics since World War II.

8
FROM COLLAPSE TO COMPETITIVENESS: CHANGING FORTUNES AFTER 1945

Recovery in the western German chemical industry was not foreordained in 1945. Clearing the rubble left by years of bombing required enormous effort, as did the repair of long-neglected machinery and equipment. Only organizational skill and patience could allow the revamping of the transportation system and the beginnings of recovery in the coal industry, providing in turn desperately needed fuel and raw materials to more or less distant chemical plants. The impediments to recovery were massive.

Certain factors worked, however, to ease the recovery process. Destruction to I.G. facilities proved in general surprisingly slight. Most damage was confined to factory buildings, and, in a large number of plants, processing equipment was intact. Those who supervised cleanup operations, repairs to plant, and resumption of production were in general men who had extensive experience in production and its management. The Hoechst plant and the other I.G. factories of the U.S. zone of occupation were exceptions to this general rule, since wartime plant managers were dismissed from their posts. But even here the men who took over the responsibility for operating the plant were generally in middle-level management within the company during the I.G. period. In addition, important governmental and business organizations (e.g., the Provincial Economic Offices and the industry trade associations) at the *Land* and local level remained intact or were rebuilt on pre-1945 lines. They were able to "administer the shortages" of the immediate postwar period and thus prevent a total collapse of the German economy. Furthermore, realization of the centrality of German recovery to western European recovery gradually eroded Allied restrictions on

German industry. Finally, the changing international political climate after World War II encouraged the western Allies to continue in this direction, and the incipient western German regime reacted by enacting economic reforms favorable to a thriving industrial environment.

The primary impetus for the recovery of the major western German chemical firms by 1950 was thus a set of forces that had their origins in the pre-1945 period, and which were to some extent channeled and directed by western Allied and German policies. Continuities—in organizational structures, managerial acumen, systems of technology and information, leading personnel, and even in physical plant—drove the recovery of the large-scale western German chemical industry. But the result was not simply a return to the antebellum status quo. These strands of continuity, and the recovery they helped bring about, formed the preconditions for the resurgence of the I.G. successors in the 1950s. The resurgence itself could, however, only occur in combination with vital changes in the bases of the German chemical economy.

Two changes in the chemical industry stand out above all others. For one thing, the dominant firm in the German chemical industry, the I.G. Farbenindustrie A.G., was replaced by three major successors of comparable size and prospects. Together, Hoechst, Bayer, and BASF were once again dominant in the West German chemical industry and extremely powerful in international markets. The three corporations could, to some degree, act together through the chemical industry trade association. But the fact that they were legally separate and consequently pursued different and competing interests marked a qualitative change in the structure of the chemical industry in Germany. Rather than tending toward monopoly, the industry was characterized by oligopoly.

The second major structural change in the large-scale West German chemical industry (and in West German industry in general) was from autarky, or self-sufficiency, to dependence. The industry, and especially the I.G. Farben, had always depended heavily upon foreign markets, and even during the Nazi period the I.G. played a central role as a foreign exchange earner. Nonetheless, there were two important ways in which Farben's production was based on the principles of autarky, especially during the Nazi period. First, representatives of the firm worked closely with the regime to supply it from within Germany with the goods needed to wage a war of expansion in the east. Without synthetic rubber and gasoline produced from German coal, water, air, and creativity, Nazi aggression would have been severely handicapped. Second, production was based on principles of autarky even before Hitler came to power in 1933 to the extent that the supply of the most important fuel and raw material for organic chemical production, coal, came from within Germany.

Political and economic pressure forced the complete rejection of the ideas of autarky after 1945. The Allies and the fact of the division of the country forced German government and industry to recognize the folly of further attempts at self-sufficiency. Commercial pressure from, above all, U.S. firms forced the I.G. successors to move during the 1950s from coal-based to petroleum-based chemistry. In the process, successors to the I.G. and many other chemical firms in West Germany became dependent upon overseas oil, supplied by American and British companies. Heavy dependence on foreign markets *and* supplies came to characterize the industry.[1]

German industrialists and representatives of the federal government should be given credit for adapting successfully to the changed conditions. In general, all have come to recognize the merits of oligopoly in the West German chemical industry. The three successors have managed to compete vigorously against one another, even in foreign markets, without major crises. Competition among the three stimulates creativity in marketing and technology but has not caused destructive behavior. This, combined with the virtually unbroken prosperity since the early 1950s, has militated against the reformation of the I.G. The growing commitment to foreign markets and supplies—in other words, the complete integration of the German economy into the capitalist world economy—justifies itself with each successive year of production and profit growth.

All three major successors to the I.G. have met with astonishing success. Each made significant additions to its holdings in the period since 1950.[2] In the 1960s, for instance, BASF's leadership sought to guarantee supplies of petrochemicals and other basic materials through acquisition of Gewerkschaft Victor Chemische Werke, a former synthetic oil facility, and Wintershall A.G., a diversified oil concern. More recently, BASF sought a foothold in the pharmaceuticals industry by purchasing Knoll A.G. in Ludwigshafen. Hoechst's management also made the effort to assure supply of raw materials by acquiring Ruhrchemie A.G., the former holder of patents for the Fischer-Tropsch synthetic oil process. Bayer, on the other hand, sought to expand its already existing production facilities. As Leverkusen finally approached the limits to expansion, the firm moved to develop further its complex at Dormagen. A purchase of land adjacent to Uerdingen in 1954 allowed further expansion of the facility, which parallelled the main plant in its breadth of production areas. In the late 1970s, Bayer's management made their first addition since the founding of I.G. Farben to their core domestic factories by opening a plant to produce preliminary products for plastic at Brunsbüttel in Schleswig-Holstein.

Each of the I.G. successors sought access to world markets as well. By far the most important in the years since 1950 were their forays into the lucrative U.S. market. BASF, after purchasing a number of smaller holdings in the United States beginning in 1958, founded the BASF Wyandotte Corporation with the acquisition of Wyandotte Chemicals Corp. in 1969. Hoechst began its inroads into the American market in 1955 with acquisition of a small dye factory. Thereafter, the Hoechst management has expanded production in the United States considerably and has brought them together into the Hoechst Celanese Corporation. Bayer's traditional mastery of foreign markets was demonstrated again already in 1954, when, together with Monsanto, Bayer's management founded Mobay Chemical Co. of Wilmington, Delaware. Later Bayer brought Monsanto's 50-percent share and Mobay was subsumed under the aegis of Bayer's holding company, Rhinechem Corporation (renamed Bayer USA Inc. in 1986). Bayer also purchased the pharmaceutical firm, Miles Laboratories Inc. of Elkhart, Indiana.

As a result of their domestic and international expansion and continued excellence in research and development, the I.G. successors moved into commanding positions in world chemical sales. Already in 1964, all three firms were ranked in the top ten chemical companies in the world: Bayer was ranked fifth; Hoechst, sixth; and BASF, tenth. Du Pont was the largest chemical company in the world, with a turnover of more than twice that of Bayer. Ten years later, BASF was the largest chemical firm in the world in terms of sales turnover. Hoechst was second, and Bayer fifth. Each had at least tripled its turnover, and the number of employees had gone up by at least 55,000, or close to 100 percent. Sales and profits continued to soar for the I.G. successors in the following ten years, although they once again surrendered the sales lead among world chemical firms to Du Pont. World sales for Bayer, Hoechst, and BASF (ranked respectively two, three, and four in the international chemical industry) went up by at least 75 percent between 1974 and 1984.[3]

The Germans adapted successfully to structural alterations in their chemical industry that occurred in the years immediately following World War II; the western Allies caused them. All of the Allies agreed from the outset that the German tendency toward autarky had to be eliminated as a precondition for a lasting peace. The emergence of a capitalist world economy provided the framework within which Germany could wean itself away from prior delusions of self-sufficiency.

Allied impact on the disentanglement of the I.G. was a more complicated matter. Initially, French and British occupation authorities moved to preserve the works groups in their respective zones and to retain the leaders of those groups in their former positions. French and British action prejudiced the eventual outcome of the disentanglement process, and to this extent one must recognize their impact on determining the structure of the West German chemical industry. The ultimate solution to the problem of how to break up the I.G. depended upon the explicit agreement of the most powerful of the occupying powers, the United States.

Two key qualities characterized the development of U.S. policy toward the I.G. First, occupation authorities argued vociferously among themselves over the best way to break up the chemical trust. Infighting within military government over the optimal and/or most desirable structure for the chemical industry in western Germany mirrored a long-standing struggle among U.S. government policymakers, and within American society itself.[4] Should a given industry be dominated by a small number of large companies, or should it be characterized by a large number of medium and small companies? U.S. agreement to break up the I.G. into three major successors and a few minor ones decided the issue in favor of the former option. The final American position on the issue of the disentanglement of the I.G. was the product of a fundamental consensus in opposition to monopoly among all groups within the U.S. military government in Germany and in American society in general, and of the relative strength of the pro-oligopoly group. In the case of the I.G. successors, and for most U.S. domestic industry groups as well, Americans have shown themselves willing to accept very large-scale firms as long as no single firm dominates an industry group.

The other major component of U.S. policy toward the I.G. was the rapid and extensive transfer of responsibility to implement policy to the Germans. Increasing autonomy in the implementation of policy involved more and more influence over the formation of policy itself. Western Germans from all camps—from government, industry, and labor—were all in agreement about the ends to which the growing influence should be applied: to the preservation of large-scale western German chemical firms that would have the technological, financial, and commercial resources to compete effectively on world markets. At the same time, representatives of government and industry in western Germany recognized from the start that they would not be permitted to form the I.G. once again.

Thus, although French and British military government policy toward the I.G. in the immediate postwar period served to constrain the options for disentangling the firm, in the final analysis the disentanglement was a German–American affair. The compromise reached—three large successors fairly equal in power and prospects along with a number of minor successors—fit in well with the German desire for a chemical industry that would be competitive on world markets. The compromise also represented an acceptable solution for U.S. policymakers to the problem of curbing the monopolistic tendencies of the German chemical industry. Once the decision was reached within the U.S. occupation regime, its representatives moved quickly to implement it. As a British commentator noted in 1953 as the very last negotiations on the I.G. were taking place:

> Having spent some years trying to prevent the American steam roller from running smoke [sic] in one direction (splinterisation, sale by auction etc.) I now find myself trying to prevent the machine from getting

out of hand in the reverse direction, because having decided to give way on matters which they have hitherto regarded as of paramount importance, they now wish to have the whole thing wound up in a matter of days and to achieve this result seem ready to accept any and every German proposal without even giving it cursory examination. . . .⁵

The forces of continuity in the western German chemical industry—in equipment, technology, personnel, and organization—were in the main responsible for its recovery. Forces of continuity, combined with changes in industry structure and in the international political climate, made for its resurgence. To this extent, 1945 at the same time was and was not a turning point for the industry. The chemical industry, though, differed from many others in a number of important respects. For one thing, its production growth rates after the war differed from those of most other industry groups. From 1962 to 1973, for instance, chemical production in the Federal Republic of Germany increased by more than three times and thus outperformed all other sectors of the economy.⁶ What is more, the chemical industry is science-based, and this too differentiates it from most other industry branches. The need to invest enormous sums in research, to gamble on new products, to search continuously for improvements in products and processes, tends to demand a risk-taking ability that separates chemical industrialists from most others. Finally, the chemical industry in Germany is of necessity among the most export-oriented.

More studies of individual industrial branches in occupied Germany, and of the postwar performance of the chemical industries in other nations, are needed to determine the extent to which the experience of the western German chemical industry was the norm. But it is clear that in several key areas, its experiences in the postwar period were typical ones. For one thing, industries across the board, whether located in a victorious or defeated nation, faced enormous changes in the postwar era. The emergence of a capitalist economic order dominated by the military and economic power of the United States was one of the most important changes. All industries in all western countries had to compete in a relatively open economy as tariff rates were systematically lowered through the General Agreement on Tariffs and Trade. At the same time, all European industrialists dealt more or less effectively with the advent of extensive European economic cooperation that involved major alterations in markets and strategic planning for business.

Unprecedented economic growth during the 1950s was a phenomenon that all industries experienced as well, although chemicals was one of the most successful sectors. For one thing, it obviated the need for price-fixing or market-division agreements among competitors: there was simply plenty of business to go around for everyone. This was obviously a break with previous European practice, especially in Germany but also in France and Britain.[7] What is more, astounding growth rates gave European chemical firms—especially those in Germany—time to recover from the war. American firms were, owing to overwhelming demands for their products at home and in closeby foreign markets, relatively uninterested in the European markets. Had they been more interested, European producers may have had a more difficult time recovering from the war. As it turned out, U.S. firms developed an extensive interest in Europe only in the late 1950s and during the 1960s, by which time the Europeans were in a position to compete more effectively.[8]

In addition, for the chemical industry in particular, new products and processes, and, even more important, new technologies based on petroleum rather than coal, had to be dealt with. Other technology-intensive industries, most notably electronics, featured similar and revolutionary developments in production and product technologies, high research costs, and innovative marketing techniques.

Furthermore, all industries emerged from the war with more or less of their production facilities intact. The decision to repair, renew, or retain existing plant was an important one. It would determine to a large degree the success or failure of corporate strategy. Finally, all industries in all countries emerged from the conflict with a certain core group of managers who had been active before or during the war. The response of this generation of managers to the problems and prospects of the postwar era determined more than anything else the performance of individual firms.

Allied—and especially American—policy provided the framework within which the resurgence of the German chemical industry could occur. But the men who ran the industry carried it out and distinguished themselves by their adaptability and creativity. The irony is that the same creativity and adaptability that allowed German industrialists to embrace autarky and to prepare Hitler's armies with the tools needed for aggression—qualities often exercised by the very same men—were responsible for the success of West German chemical manufacturing under the new conditions of the postwar period.

NOTES

INTRODUCTION

1. Recently scholars have begun to consider this problem by concentrating chiefly on heavy industry. See Alan Milward, *The Reconstruction of Western Europe, 1945–1951* (London: Methuen, 1984); Volker Berghahn, *The Americanisation of West German Industry, 1945–1973* (Cambridge/New York: Cambridge University Press, 1986).

2. One indication of this is the industry's production in comparison to the coal industry: the West German coal industry only reached its 1938 production level again in 1955, at the same time that production in the country's chemical industry had already soared to 181 percent of the 1938 level. Verband der Chemischen Industrie, e.V., *Chemiewirtschaft in Zahlen*, 4. Auflage (Düsseldorf: Econ Verlag, 1960), pp. 38–39. For articles that analyze the performance of West German chemical firms in the 1950s and 1960s in the international context, see, for instance, Gilbert Burck, "Chemicals: The Reluctant Competitors," *Fortune* 68 (November 1963): 148–153, 218–228; and, John Davenport, "The Chemical Industry Pushes into Hostile Territory," *Fortune* 79 (April 1969): 108–114, 156–162.

1. RISE AND PRECIPITOUS FALL OF GERMAN *GROSSCHEMIE*

1. Victor Lefebure, *The Riddle of the Rhine: Chemical Strategy in Peace and War* (New York: The Chemical Foundation, 1923), p. 18. Note that the I.G. that Lefebure wrote about was a cartel, and that it became a single corporation in 1925. See below in this chapter.

2. Helmuth Wickel, *I.G.-Deutschland: Ein Staat im Staate* (Berlin: Verlag Der Bücherkreis, 1932), p. 213.

3. Joseph Borkin, *The Crime and Punishment of I.G. Farben* (New York: Free Press, 1978; Pocket books ed., 1979), quotes from pp. 3 and 198. Borkin's book is more lawyer's brief than scholarly monograph; he often skews documentary information to demonstrate his points.

4. Peter Wolfram Schreiber (Pseudonym of Autorenkollektiv von

Mitgliedern der Kommunistischen Studentengruppen), *I.G. Farben—Die unschuldige Kriegsplaner* (Stuttgart: Verlag Neuer Weg, 1978), quote from cover.

5. L. F. Haber, *The Chemical Industry in the Nineteenth Century* (Oxford: Clarendon Press, 1958), pp. 39–50.

6. Ibid., pp. 63 ff., 129.

7. Jürgen Kocka and Hannes Siegrist, "Die hundert grössten deutschen Unternehmen im späten 19. und frühen 20. Jahrhundert. Expansion, Diversification und Integration im internationalen Vergleich," in Norbert Horn and Jürgen Kocka, eds., *Recht und Entwicklung der Grossunternehmen im 19. und 20. Jahrhundert* (Göttingen: Vandenhoeck & Ruprecht, 1979), pp. 103–104, 111.

8. Haber, *Nineteenth Century*, p. 129.

9. Table in Bayerwerksarchiv, Leverkusen (hereafter BWA), Produktionsbeginn: Unterlagen 1945 (o. Titel).

10. Ibid. Material on patent law from Paul M. Hohenberg, *Chemicals in Western Europe: 1850–1914. An Economic Study of Technical Change* (Chicago: Rand McNally, 1967), pp. 71 ff. Quotation from p. 72.

11. Wolfram Fischer makes a similar point for German industry in general in his essay on "The Role of Science and Technology in the Economic Development of Modern Germany," pp. 71–113 in William Beranek, Jr., and Gustav Ranis, eds., *Science, Technology, and Economic Development: A Historical and Comparative Study* (New York: Praeger, 1978). See p. 79.

12. W. G. Hoffmann, F. Grumbach, H. Hesse, *Das Wachstum der deutschen Wirtschaft seit der Mitte des 19en Jahrhunderts* (Berlin: Springer, 1965), p. 63. It is possible that there is a base-line problem with these figures; that is, the chemical industry started at a relatively low level in 1872 and its subsequent gains may appear more impressive than they were for that reason. Still, I think they are reasonably accurate: the figure for all industry includes both established industrial sectors already producing at a high rate and newer branches that started at a low rate of production. Thus, the two sets of developments tend to cancel one another out in the summary statistics. In addition, the chemical industry had already grown rapidly in the 1850s and 1860s, and thus was not starting out "near the base line."

13. Graham Taylor and Patricia E. Sudnick, in *Du Pont and the International Chemical Industry* (Boston: Twayne, 1984), point out that ties between German chemical firms and banks were stronger than parallel cases in other countries (p. 12). But their characterization is somewhat misleading since larger firms did not depend so heavily on banks, especially as the industry matured. For the later experience of I.G. Farben in the financial

arena, see Peter Hayes, *Industry and Ideology: I.G. Farben in the Nazi Era* (Cambridge/New York: Cambridge University Press, 1987), p. 27. For the industry's dividend rate, ibid., p. 7. The implications of the lack of dependence on banks are explored in greater detail in chapter 7.

14. Gottfried Plumpe, "The I.G. Farben Group as a Multinational Enterprise between the Two World Wars," European University Institute, Florence: EUI Colloquium Papers, "The Early Phase of Multinational Enterprise in Germany, France, and Italy," Florence, 17–19 October 1984 (1984), pp. 10–25, esp. pp. 24–25. Plumpe's case that Farben was a multinational corporation is overstated.

15. For further information on Duisberg and his ideas and accomplishments, see: chapter 3 below; Hans-Joachim Flechtner, *Carl Duisberg. Vom Chemiker zum Wirtschaftsführer* (Düsseldorf: Econ Verlag, 1959); and the entry on Duisberg in *Neue deutsche Biographie*, Bd. 4 (Berlin: Duncker & Humblot, 1959), pp. 181–182.

16. L. F. Haber, *The Chemical Industry 1900–1930: International Growth and Technological Change* (Oxford: Clarendon Press, 1971), pp. 108–109.

17. For a particularly good synopsis of the program, its origins, and its implications, see Haber, *1900–1930*, pp. 198 ff.

18. Haber, *Poisonous Cloud*, passim, esp. pp. 171 and 244; Lefebure, *Riddle*, passim, esp. pp. 17, 36, 239–240; on the attack at Ypres and its implications, see Ulrich Trumpener, "The Road to Ypres: The Beginnings of Gas Warfare in World War I," *Journal of Modern History* 47 (September 1975): 460–480.

19. Lefebure, *Riddle*, p. 17. L. F. Haber, in his more recent, and definitive, study of gas warfare in World War I, appears to differ with Lefebure's estimate. Haber claims instead that "in 1918 . . . 28 per cent of all German shells contained chemical warfare agents." Of course, this does not preclude the possibility that the percentage of shells filled with gas used in the retreat was higher. Moreover, Haber admits that this was "a proportion unsurpassed by any other belligerent." See Haber, *The Poisonous Cloud: Chemical Warfare in the First World War* (Oxford: Clarendon Press, 1986), pp. 260–261. Neither Lefebure nor Haber draws the conclusion noted here from the high percentage of gas shells used in the retreat. I owe this insight to a conversation with Professor Alan Beyerchen.

20. Germany's share of world chemical production dropped significantly as a result of the war. In 1913, Germany produced over one-fourth of total world chemical products, while in 1927 that percentage had dropped to just under 17 percent. See Alfred Maizels, *Industrial Growth and World Trade* (Cambridge: Cambridge University Press, 1963), p. 292, table 11–4. On Du Pont, see Taylor and Sudnick, *Du Pont and the International Chemical Industry*, pp. 59–74. The book gives a useful overview of internal

developments at Dupont and the company's place in the international chemical industry from the early nineteenth century to the present.

21. W. Treue, "Carl Duisbergs Denkschrift von 1915 zur Gründung der 'Kleinen I.G.,'" *Tradition* 8 (1963): 205–206. Treue's article is a reproduction of the actual memo, with a brief introduction to it. The synopsis here is drawn from Duisberg's own analysis.

22. Ibid., p. 201.
23. Ibid., pp. 202–204.
24. Ibid., p. 207.
25. Ibid., pp. 210, 218–219.
26. Ibid., p. 207 ff.
27. Ibid., pp. 199 and 215 ff.
28. The firms that would make up the I.G. had mixed experiences during the inflation period. They were affected by uncertainty, especially during the hyperinflation. But they also were insulated from some of the worst effects of inflation by virtue of their foreign sales and subsidiaries. They took advantage, in addition, of the erosion of the mark's value to pay back government loans from the war period, especially those that had helped finance expansion of synthetic nitrogen facilities. See Hayes, *Industry and Ideology*, pp. 12–13. On the inflation in its wider context, see the excellent synthesis by Carl-Ludwig Holtfrerich, *The German Inflation, 1914–1923* (Berlin/New York: de Gruyter, 1986).

29. Calculated from figures in Hermann Gross, *Material zur Aufteilung der I.G. Farbenindustrie Aktiengesellschaft* (Kiel: Institut für Weltwirtschaft, 1950), Tabelle Ia.

30. Calculated from figures in ibid. I have defined net profit as distributed profits (dividends) plus retained earnings. For more on the I.G.'s profit picture, see later in this chapter.

31. For a thorough analysis of this, see Peter Hayes, "Carl Bosch and Carl Krauch: Chemistry and the Political Economy of Germany, 1925–1945," *Journal of Economic History* 47 (June 1987): 354–363.

32. Gross, p. 7.
33. Ibid., analysis based on figures in Schaubild Vb.
34. Ibid., p. 27; and Tabelle IVa. The comparisons with Deutsche Solvay-Werke are drawn from tables in Haber, *1900–1930*, pp. 290–291. Note that "RM" refers to *Reichsmark*, the German currency until June 1948, when it was replaced in West Germany by the *deutsche mark* (DM).

35. Gross, *Material*, p. 11.
36. Figures drawn from Dr. Ernst Struss, Erklärung unter Eid, 12.VI.1947 (Übersetzung des Dokumentes Nr. NI-7236, Office of Chief of Counsel for War Crimes), Bundesarchiv [hereafter BA] Koblenz, All.

Proz. 1/IX B 46. They agree generally with those presented by Hayes in *Industry and Ideology*, pp. 17–18.

37. Hayes, p. 18.

38. There is a long list of the foreign subsidiaries of the I.G. in OMGUS, *Ermittlungen gegen die I.G. Farben* (a set of German translations of English-language documents) (Nördlingen: Franz Greno, 1986), pp. 270–293. Sales abroad were mainly in the areas of pharmaceuticals, dyes, photographic materials, and high-technology licensing and cooperation (ibid., p. 23), that is, in the company's strong areas. For a more detailed discussion of this problem, see chapter 2.

39. It is not possible here to go into extensive detail on the organization of the firm. Interested readers may consult Gross, *Material*, pp. 7–10; Helmuth Tammen, *Die I.G. Farbenindustrie A.G. 1925–1933* (Berlin: Verlag Helmuth Tammen, 1978), pp. 21–28; *Trials of War Criminals*, Vol. VII (Washington: U.S. Government Printing Office, 1953), pp. 378–413.

40. Tammen, *I.G. Farben*, pp. 27–28.

41. Tammen, *I.G. Farben*, pp. 27–28. See also extracts of testimony by von Knierheim (p. 389), and Ilgner (pp. 390–392) in *Trials*, Vol. VII.

42. Tammen, *I.G. Farben*, pp. 24–25.

43. Ibid., pp. 23–24.

44. This figure is based on the list presented in ibid., figure facing p. 28. The I.G. had participations in hundreds of foreign and domestic firms and ran other facilities in Germany under contract from the government. Lists of foreign and domestic holdings of the I.G. in 1945 appear in OMGUS, *Ermittlungen*, pp. 270–306.

45. Tammen, p. 21.

46. Ibid., p. 22.

47. See chapter 7.

48. For more on the technological interconnections among plants in each of the three major production groups in western Germany, see chapters 2–4.

49. The figures were compiled for the use of British military government in Gemany and probably originated from questioning I.G. officials. Peter Hayes has suggested to me that the figures of Auschwitz are almost certainly too low.

50. On the choice of Auschwitz, see, for instance, Raul Hilberg, *The Destruction of the European Jews* (Chicago: Quadrangle, 1971; Quadrangle Paperback ed., 1967), pp. 592–595. More recent, and more firmly grounded in the concern's own records, are the accounts by Peter Hayes, *Industry and Ideology*, pp. 347–351; and Peter J. T. Morris, "The Development of Acetylene Chemistry and Synthetic Rubber by I.G. Farbenin-

dustrie Aktiengesellschaft, 1926-1945" (Dissertation, Oxford University, 1982), pp. 330-345.

51. On synthetic fuels development and the I.G., see especially Hayes, *Industry*, pp. 36-42, 115-119, 133-135, and 183; Wolfgang Birkenfeld, *Der synthetische Treibstoff 1933-1945* (Göttingen: Musterschmidt, 1964); Arnold Krammer, "Fueling the Third Reich," *Technology and Culture* 19 (July 1978): 394-422; Raymond G. Stokes, "The Oil Industry in Nazi Germany, 1936-1945," *Business History Review* 59 (Summer 1985): 254-277.

52. See for instance Borkin, *Crime and Punishment*; Richard Sasuly, *I.G. Farben* (New York: Boni & Gaer, 1947).

53. Thomas P. Hughes, "Technological Momentum in History: Hydrogenation in Germany 1898-1933," *Past and Present* 44 (August 1969), pp. 108-120.

54. The meeting between Hitler and Gattineau and Bütefisch is central to understanding the relationship between party and firm in the pre-1933 period according to the accounts of Borkin, *Crime and Punishment*, pp. 69-70; Sasuly, *I.G.*, p. 110.

55. Borkin, p. 71; Sasuly, pp. 62-71.

56. Hayes, *Industry*, pp. 66-67; Henry A. Turner, *German Big Business and the Rise of Hitler* (New York: Oxford University Press, 1985), pp. 246 ff. (Turner relies heavily on Hayes's dissertation, "The *Gleichschaltung* of I.G. Farben" [Dissertation, Yale University, 1982]); Heinrich Gattineau, *Durch die Klippen des 20. Jahrhunderts* (Stuttgart: Seewald Verlag, 1983), pp. 127-128.

57. Hayes, pp. 61-64; Turner, pp. 235, 260 ff.

58. Hayes, pp. 67-68; 158-161; Turner, pp. 246 ff.

59. This argument formed the basis for U.S. Military Government General Order No. 2 Pursuant to Military Government Law No. 52 (Blocking and Control of Property): I.G. Farbenindustrie A.G., July 5, 1945. In Hoechst A.G., *Dokumente aus Hoechst Archiven*, vol. 48: "U.S. Administration. Die Verwaltung des Werkes Hoechst 1945-1953" (Hoechst: Hoechst A.G., 1976), p. 30.

60. This is the argument and title of the essay by Timothy Mason, "The Primacy of Politics: Politics and Economics in National Socialist Germany," pp. 165-195 of S. J. Woolf, ed., *The Nature of Fascism* (New York: Random House, 1968).

61. Peter Morris's contention (in "Development of Acetylene Chemistry," pp. 77-82) that I.G.'s net profits were stunningly low in the Nazi period is misleading. His figures for net profits appear to be based solely on dividends paid out by the company and do not take into consideration retained earnings, research expenditures, and so on. The numbers I present in the following are conservative estimates of the firm's profits after

taxes, including only dividends and retained earnings. I do not include research expenditures or tax write-offs.

Inflation is not taken into consideration in the figures presented, even though it eroded the value of money substantially by 1944. Since profits are presented as a percent of turnover, this problem is less significant.

62. Calculated from Gross, *Material*, Tabelle Ia. I.G.'s pattern of relative neglect of basic research during the war years was also characteristic of government-sponsored weapons research efforts in the same period. See Leslie Simon, *German Research in World War II* (New York: John Wiley, 1947), p. 207.

63. Gross, ibid. See chapter 7 for further analysis of dividend policies during the I.G. period.

64. See Hayes, *Industry*, chaps. 6 and 7; on the occupation of France in particular, see Rosemarie Denzel, *Die chemische Industrie Frankreichs unter der deutschen Besetzung im zweiten Weltkrieg* (Tübingen: Institut für Besatzungsfragen, 1959) (mimeographed, available in Institut für Zeitgeschichte in Munich), esp. pp. 139–140.

65. Much of the story of I.G.'s involvement in synthetic fuels is told well in Hayes, *Industry*, pp. 36–42, 115–119, 133–135, and 183. Hayes makes the point that its involvement in synthetic fuels development was disappointing more explicitly in "Carl Bosch and Carl Krauch," pp. 359–360. My thanks to Professor Hayes for providing me with a manuscript of his paper prior to publication.

66. For further information on these and other cases, see Hayes, *Industry*, pp. 156–158, 176–180, 337–339, and 374; and the biographies of the defendants in the I.G. Nuremberg trial, in *Trials of the War Criminals* 7: 338–391, 425, 435, 544 (Bütefisch), 607 (Krauch), 614, 619, 633, 642, 1145, 1149, 1334, and 1498, and 8: 787. The information on Bütefisch in the *Freundeskreis* is in Louis Lochner, *Tycoons and Tyrant* (Chicago: Henry Regnery, 1954), p. 108, as well. The *Freundeskreis*, successor to an organization formed in early 1932 around Keppler, advised the Party on economic affairs and served to link the business community with the Nazi Party.

67. Peter Hayes tells the story of I.G. at Auschwitz in chap. 8 of his *Industry*, pp. 347–368.

68. Hayes, *Industry*, pp. 158, 176–180. Hayes qualifies this argument elsewhere in his book, implying at one point that Krauch represented the interests of the I.G. adequately at least in planning in the Four Year Plan organization. See ibid., pp. 206–207. Later he points out that "Krauch may not have been Farben's servant in high office, but he and his staff at least provided access to influence." See ibid., pp. 372.

69. Hayes deals to some extent with the I.G. employees involved in the

regime, but detailed information on them—including numbers, background, and influence—appears to be scanty. See Hayes, pp. 177-180, 338-339, and 374. He deals briefly with the case of Bütefisch in chap. 8, pp. 338-339.

70. For this argument in general, see Timothy Mason, "The Primacy of Politics"; on the oil industry in particular, see Stokes, "The Oil Industry in Nazi Germany."

71. Hayes, *Industry*, pp. 183-188, esp. p. 185.

72. Mason, "Primacy of Politics"; Hayes, *Industry*, p. 163; R. J. Overy, "Heavy Industry and the State in Nazi Germany: The Reichswerke Crisis," *European History Quarterly* 15 (July 1985): 313-340; Overy, "Hitler's War and the German Economy: A Reinterpretation," *Economic History Review* 35 (May 1982): 272-291; on autarky, see William Carr, *Arms, Autarky, and Aggression* (New York: W. W. Norton, 1972).

73. Hayes, *Industry*, chap. 7. The following draws heavily upon Hayes's analysis in part IV, chaps. 7-8, of his book. On German industry's postwar planning in general, see especially Robert Herzstein, *When Nazi Dreams Come True* (London: Abacus, 1982). Ludolf Herbst also deals extensively with this subject in *Der totale Krieg und die Ordnung der Wirtschaft* (Stuttgart: Deutsche Verlags-Anstalt, 1982).

74. Hayes, *Industry*, pp. 266-271.

75. Wolfgang Heintzeler, *Der rote Faden* (Stuttgart: Seewald, 1983), p. 114; Paul Danek, "Zur reaktionären Rolle des staatsmonopolistischen Kapitalismus bei der Wiedererrichtung und Machtausweitung des I.G.-Farbenmonopols in Westdeutschland" (Dissertation, University of Halle, 1961), p. 104; Hermann Schreyer, "Der I.G.-Farben-Konzern, seine Vorgänger und Nachfolger. Ein Beitrag zur Organisationsgeschichte der deutschen chemischen Industrie," *Archivmitteilungen* 16 (1966), p. 157; Tammen, *I.G. Farben*, p. 26; Jürgen Räuschel, *Die BASF. Zur Anatomie eines multinationalen Konzerns* (Köln: Pahl-Ruggenstein, 1975), p. 21.

76. Interview of author with Prof. Dr. Kurt Hansen, Leverkusen, 4 December 1984.

77. Direction Politique Europe, M. Jacques Truelle (Ministre Plenipotentiaire délégué en Espagne du Gouvernement Provisoire de la Republique Française) to M. Georges Bidault (Ministre des Affaires Etrangères), March 7, 1945, and attached report from March 1, 1945—French Foreign Ministry Archives, Paris, Série Z, 89, pp. 22-23.

78. Hayes, *Industry and Ideology*, pp. 369-370, 375-376, for other examples of limited postwar planning. My thanks to Professor Hayes for expanding on this in a personal conversation.

Interestingly enough, this tactical preparation for projected Allied breakup of the firm appears to have been in keeping with Schmitz's general style of managing. Josiah DuBois, Jr., who later led the prosecution of I.G. executives in Nuremberg, claims that Schmitz's file in Berlin NW7 indicated that Schmitz had pursued a new course upon assuming the post of head of the managing board in 1935. Until then, all factories operated by Farben carried its name. Thereafter, those plants producing peacetime and/or wartime goods carried the I.G. name and were self-financed with guarantees against losses from the Reich. Those producing war-related goods *only* (e.g., explosives, military gases, etc.) were simply operated by the I.G., did not carry its name, and were financed entirely by the *Wehrmacht* or other Reich authorities. See DuBois, *Generals in Grey Suits* (London: The Bodley Head, 1953), p. 47. Hayes alludes to this practice as well on p. 186.

79. United States Strategic Bombing Survey (European War), USSBS Rpt. 109, *Oil Division Final Report* (2d. ed., Jan. 1947), p. 1.

80. For an analysis of the effects of the bombing on production in the chemical industry, see USSBS 109, pp. 40–47, 101–111, and appendix A, pp. 1–20.

2. RADICAL DECONCENTRATION AND ITS RESULTS

1. There are numerous versions of the characterization of the zones of occupation. Douglas Botting, for instance, notes that "The Russians . . . got the corn, the French the wines, the British the ruins and the Americans the scenery." See Botting, *In the Ruins of the Reich* (London: George Allen & Unwin, 1985), p. 153. Clarence Lasby, in *Operation Paperclip: German Scientists and the Cold War* (New York: Atheneum, 1971), has a slightly different version from the one presented here, but more importantly he notes the "scientific bonanza" available in, and extensive wartime removals of physical plant, to the U.S. zone, which undermines the contention that all the United States got was the scenery. See Lasby, p. 49.

2. Percentages calculated on the basis of figures in Industriestatistik-Archivmaterial, "Netto-Produktionswerte der deutschen Industrie 1936," n.d., Bundesarchiv (hereafter BA), Koblenz Zl/870. Bremen is not included in the U.S. zone calculations.

3. OMGH Statistical, Historical, and Field Reports Division, "Subject: Industrial Structure of the Combined Area in Comparison to Hesse, 21 September 1948," Hauptstaatsarchiv (hereafter HStA), Hessen, Wiesbaden, 649, OMGH Historical Division, 8/189/5.

4. An indication of the financial resources in the U.S. zone is found in the discussion on providing the BASF plants in the French zone with operating capital in, Allied Control Authority (ACA), I.G. Farben Con-

trol Committee, "Minutes of the First Meeting Held in Berlin on 4 February 1946," COIG/M(46)1, 25.II.1946, Public Record Office, London (Kew) (hereafter PRO), T236/975. According to a military government report, the United States seized assets at the central office of the I.G. on July 5, 1945, including bank accounts of about RM 86 million, as well as about RM 89 million in notes. See OMGUS, "Control of I.G. Farben: Special Report of the Military Governor" (n.p., October 1, 1945), p. 24. A summary of the state of the I.G.'s written records in the U.S. zone at the end of the war is in Karl Heinz Roth's introduction to OMGUS, *Ermittlungen gegen die I.G. Farbenindustrie AG—September 1945* (Nördlingen: Franz Greno, 1986), pp. xiii–xxiv.

5. The historical literature on the U.S. zone and American policy developed early and is much more extensive than that for the other western zones. The classic general work on the occupation is John Gimbel, *The American Occupation of Germany* (Stanford: Stanford University Press, 1968), but the literature on the subject is vast. See also Edward N. Peterson, *The American Occupation of Germany: Retreat to Victory* (Detroit: Wayne State University Press, 1977); and the extended consideration of U.S. policy by Hans-Peter Schwarz in *Vom Reich zur Bundesrepublik: Deutschland im Widerstreit der außenpolitischen Konzeptionen in den Jahren des Besatzungsherrschaft 1945–1949*, 2d enl. ed. (Stuttgart: Klett-Cotta, 1980), esp. pp. 39–146. To this point, most studies have not been concerned with specific industry branches.

6. See, for example, Alfred E. Eckes, Jr., *A Search for Solvency: Bretton Woods and the International Monetary System, 1941–1971* (Austin/London: University of Texas Press, 1975); or, more generally, John Lewis Gaddis, *The United States and the Origins of the Cold War, 1941–1947* (New York: Columbia University Press, 1972).

7. For Treasury influence, see Eckes; also see Daniel Yergin, *Shattered Peace* (Boston: Houghton-Mifflin, 1978). A discussion of talks within the administration on German policy is in Walter Dorn, "The Debate over American Occupation Policy in Germany in 1944–1945," *Political Science Quarterly* 72 (December 1957): 481–501. The Plan itself is outlined in U.S. Senate, Committee on the Judiciary, *Morgenthau Diary (Germany)*, 2 vols. (Washington: USGPO, 1967), I: 548–555 (hereafter cited as *Diary*). See also Warren Kimball, *Swords or Plowshares? The Morgenthau Plan for Defeated Nazi Germany* (Philadelphia: J. B. Lippincott, 1976).

8. Transcripts of the meetings are in *Diary*, II, passim.

9. JCS 1067/6, 25 April 1945; JCS 1067/8, 10 May 1945; JCS 1067/9, 11 May 1945, in U.S. National Archives (hereafter NA), Washington, Record Group 218 (U.S. Joint Chiefs of Staff), CCS 383.21 Germany

(2-22-44), section 8. JCS 1067/8 and 9 were amendments to the basic occupation directive (i.e., 1067/6) allowing short-term use for the purposes of the occupation of prohibited industries in Germany, particularly synthetic oil and synthetic rubber.

10. Ibid.

11. Ibid. For a detailed discussion of research control efforts in the U.S. occupation, see Alan Beyerchen, "German Scientists and Research Institutions in Allied Occupation Policy," *History of Education Quarterly* 22 (Fall 1982): 289–299.

12. "Accord on Treatment of German-Owned Patents," U.S. Department of State, *Treaties and Other International Acts Series*, No. 2415, pp. 1–3.

13. Werner Colrausz, "Vorschläge für das Patentgesetz," *Angew. Chemie* 19, Teil B (February 1947): 50. The article lays heavy emphasis on the virtues of forming a new German patent office. The following owes much to discussions in a *Kaffeerunde* at the Institut für Europäische Geschichte on June 12, 1985. My thanks to all of the participants, and in particular to Dr. Ulrich Wengenroth.

14. Bayer A.G., *Bayer in Wirtschaft und Gesellschaft* (Leverkusen, ca. 1976), p. 5.

15. As Gilbert Burck noted in 1963, "Today probably more than half the [European chemical] industry's revenues come from products that did not exist twenty-five years ago." Burck, "Chemicals: The Reluctant Competitors," *Fortune* 68 (November 1963): 148. In the same article, and writing about Bayer in particular, Burck noted that "nearly 60 percent of the company's world sales . . . now come from products that did not exist fifteen years ago." Ibid., p. 152.

16. Research in this area has been sparse thus far. Secondary literature has concentrated for the most part upon German scientists captured after the war. See Samuel Goudsmit, *Alsos*, reprint with new intro. by R. V. Jones (New York: Henry Schuman, 1947; Los Angeles: Tomash, 1983); Clarence Lasby, *Operation Paperclip*. There is very little material directly on technical exploitation. Leslie E. Simon summarizes some of the findings of the technical investigators in *German Research in World War II* (New York: John Wiley, 1947); and Lasby comments to some extent on their activities and reports, e.g., pp. 18–26, 129–130, and 163–164. More recent is an article by Arnold Krammer on the Technical Oil Mission. See Krammer, "Technology Transfer as War Booty: The U.S. Technical Oil Mission to Europe, 1945," *Technology and Culture* 22 (January 1981): 68–103.

17. Alphabetical Subject Index of CIOS, BIOS, FIAT, and JIOA Final Reports, n.d. PRO, London, FO 1005/1602.

18. See, among others, Great Britain, British Intelligence Objectives

Subcommittee, BIOS Report 1697, *Synthetic Oil Production in Germany: Interrogation of Dr. Bütefisch* (London: HMSO, n.d.); and also Krammer, "Technology Transfer."

19. CIOS File XX-II, P. J. Leaper, Reporter, "I.G. Farbenindustrie A.G. Plant Hoechst. Report on Visit 10-11/4/1945," April 25, 1945, pp. 4-6, HStA Hessen, Wiesbaden, 649, OMGH Bipartite Liaison Div., Econ. Br., 17/164-1/12.

20. FIAT Special Echelon, Special Echelon CCG (BE), "Minutes of a meeting to consider the I.G. Farben investigation in . . . Hoechst," August 10, 1945, PRO 1039/773.

21. The microfilms are available at the U.S. Library of Congress.

22. See Krammer, "Technology Transfer"; and Richard Vietor, "The Synthetic Liquid Fuels Program: Energy Politics in the Truman Era," *Business History Review* 54 (Spring 1980): 1-34.

23. Graham D. Taylor, "The Rise and Fall of Antitrust in Occupied Germany, 1945-1948," *Prologue* 11 (1979): 23-29. See also Volker Berghahn, *The Americanisation of West German Industry, 1945-1973* (Cambridge/New York: Cambridge University Press, 1986); and more generally Ellis Hawley, *The New Deal and the Problem of Monopoly* (Princeton: Princeton University Press, 1966).

24. Joseph Borkin, *The Crime and Punishment of I.G. Farben* (New York: Free Press, 1978; Pocket Books ed., 1979), p. ix.

25. James S. Martin, *All Honorable Men* (Boston: Little, Brown, 1950); Randolph H. Newman was also involved in the I.G. Farben case in Nürnberg, as evidenced in the document in Nürnberg records signed on June 5, 1947 (NI-8581—Nürnberg Prosecution, IfZ München).

26. Joseph Borkin and Charles A. Welsh, *Germany's Master Plan: The Story of Industrial Offensive* (New York: Duell, Sloan and Pearce, 1943).

27. Borkin and Welsh, ibid., pp. 7-9, 14-18, 21, 316. The book is concerned with German firms other than the I.G. (e.g., Krupp). Still, about 270 of its 316 pages are devoted to exposing the role of the I.G. On p. 21 the authors contend that "without I.G. Germany could not, twice within a generation, have filled the vials of wrath and hurled their Prussic acid in the face of the world."

28. Bernard Bernstein, who wrote one of the first policy statements on the I.G., had been assistant general counsel in the Treasury Department in the years before 1941. His career is described briefly in Josiah E. Dubois, Jr., *Generals in Grey Suits* (London: The Bodley Head, 1953), pp. 14-15; and in the introduction to *Ermittlungen gegen die I.G. Farben*. Bernstein's report has been translated into German and published in full in *Ermittlungen*, pp. 1-264.

29. Documents are interspersed throughout the files of the Decarteliza-

tion Branch in the OMGUS and OMGH records. British military government collections of the documents are in PRO FO 1031/134–136, 231–234, and FO 1013/2477.

30. See DuBois, *Generals in Grey Suits*. DuBois himself was the chief prosecutor in the I.G. trials at Nuremberg, and was associated (as counsel and consultant) with the Treasury Department before the during the war.

31. There are materials on this in PRO FO 1031/134–136. See also Borkin and Welsh, *Master Plan*, esp. pp. 21, 42–48, 53–58.

32. I.G.'s subsidiary, Degesch, did not manufacture Zyklon B, but rather coordinated its manufacture and distribution by licensed companies. For more on Degesch, see Peter Hayes, *Industry and Ideology: I.G. Farben in the Nazi Era* (New York: Cambridge University Press, 1987), pp. 361–363.

33. Ibid. See also Bernstein report in *Ermittlungen*. The real and alleged crimes of the I.G. Farben leadership are discussed to some extent in chapter 1. For more on the Nuremberg trials beginning in 1947, see the discussion at the end of this chapter, and in chapter 6.

34. See chapter 1 above.

35. OMGUS, *Ermittlungen*, pp. 23, 270–293. The main foreign holdings of the I.G. were in pharmaceuticals, dyes, and photographic materials sales and in high-technology licensing and cooperation.

36. The I.G. Farben control officer continued to exert a great influence within the U.S. military government into the 1950s, and the Decartelization Branch appears to have been quite active in trying to block approval of a report by three Allied experts in 1950 on the breakup of the I.G. into a small number of large units. C. F. McFarlane (FO German Commercial Relations and Industry Dept.), "Subject: Meeting with Mr. Brearley on 6 June, 7 June 1950," PRO FO 371/85679. Also R. J. B. Williams (Industrial Control and Decartelization Branch of the British Economic Adviser) to R. S. Swann (FO Ger. Comm. Rel. and Ind. Dept.), "Subject: I.G. Farben, 21 August 1950," PRO FO 371/85681.

37. See, for instance, the articles by H. Radandt, H. Etzold, and D. Eichholtz in *Jahrbuch für Wirtschaftsgeschichte* (1966), Teil III. The I.G. and its executives play a major role in David Abraham's description of *The Collapse of the Weimar Republic* (Princeton: Princeton University Press, 1981), although his research methodology has been successfully questioned. The polemical literature on Abraham's book has grown very large indeed and extends to articles in the *New York Times* and *Newsweek*. One of the most complete summaries of the charges and countercharges appears in *Central European History* 17 (June/September 1984): 159–293. The book has now appeared in an extensively revised edition (New York: Holmes and Maier, 1986), although the debate continues: see the review of the

new edition by V. R. Berghahn in the *New York Times Book Review* (August 2, 1987) and the replies to the review by Henry Ashby Turner, Jr., and Gerald Feldman in ibid. (September 13, 1987).

38. For Thurman Arnold's views on this subject, see Ellis W. Hawley, *The New Deal and the Problem of Monopoly*, p. 428. In the conclusion to his excellent study, Hawley deals in general with the ambivalence of the American antitrust position. See pp. 472–494. Gene M. Gressley, in "Thurman Arnold, Antitrust and the New Deal," *Business History Review* 38 (Summer 1964), notes that Arnold was not opposed to industrialism, big business, or the U.S. capitalist political–economic system per se (pp. 229–230). In a book covering a later period in the history of the Antitrust Division, Theodore Kovaleff goes so far as to claim, "that antitrust laws had been for the twentieth-century American an expression of one of the country's goals: economic freedom in a democratic society. The underlying ideology was based on the American principle of limited power, which was not only the cornerstone of the governmental system, but also the foundation of American business." "Antitrust, thus, was pro-business."

See Kovaleff, *Business and Government during the Eisenhower Administration: A study of the Antitrust Division of the Justice Department* (Athens, OH: Ohio University Press, 1980), pp. 156–157.

39. Members of several of the teams, including those for chemicals and liquid fuels and lubricants are listed in "U.S. Gets Secrets of Reich Industry," *New York Times* (July 29, 1945). For the factory level (at Vereinigte Glanzstoffwerke) see Ludwig Vaubel, *Zusammenbruch und Wiederaufbau. Ein Tagebuch aus der Wirtschaft* (Munich: R. Oldenbourg, 1984), esp. pp. 45 and 53. For the upper levels of U.S. government and military government, see Martin, *All Honorable Men*, p. 41.

40. Martin, p. 264.

41. On the admiration of German managers, techniques of management, and working habits, see for example the portion of the letter of Alfred Sloan of GM to Bernard Baruch (November 30, 1945, in the Baruch Papers), quoted in Lloyd Gardner, "America and the Geman Problem, 1945–1948," in Barton Bernstein, ed., *Politics and Policies of the Truman Administration* (Chicago: Quadrangle Books, 1970), p. 129; for the same evidence of respect and the argument for a return to business as usual, see Lewis Brown (from Johns-Manville Corp.), *Report on Germany* (New York: Farrar, Straus, 1947), esp. pp. 37–38, 67, and 83–84.

42. Martin, *All Honorable Men*, pp. 173 ff.

43. William H. Draper to Industry Branch (Wilkinson), "Subject: Penicillin, 19 August 1946," BA Koblenz, OMGUS ED Dir. Off., 3/129–1/12. It is not clear if Draper knew of the I.G.'s 50-percent participation

Notes to pages 48–51

in Dr. Alexander Wacker GmbH. (See OMGUS, *Ermittlungen*, p. 304.)

44. The exhibition opened on October 16, 1946. For a description of the negotiations leading up to it and the exhibition itself, see Karl Geiler, *Geistige Freiheit und soziale Gerechtigkeit im neuen Deutschland* (Wiesbaden: Verlag der Greif, 1947), pp. 159–165.

45. Olav Maseng (Dep. Dir. ED Hesse) to Minister of Economics and Transport for *Land* Greater Hesse, "Subject: Participation of Former I.G. Farben Plants in Wiesbaden Exhibition, 27 August 1946," HStA Hessen, Wiesbaden, 649, OMGH ED, Office of the Director, 8/80–1/19.

46. OMGUS ED Dec. Br. (by James Martin), "Report to the Economic Directorate and the Control Officers as to Seizure of Properties of I.G. Farbenindustrie A.G. in the US Zone, 15 January 1946," BA Koblenz, OMGUS ED Dir. Off., 3/127–3/8.

47. CORC, "Quadripartite Control of Plants and Assets of I.G. Farbenindustrie A.G.," Memo by U.S. Member, October 5, 1945, PRO 1039/773.

48. The following is compiled from various sources, including Farbwerke Hoechst A.G., "Farbwerke Hoechst AG," pp. A625–A640 in the special issue devoted "Zum 75.-jährigen Bestehen des Chemie-Verbandes," *Chemische Industrie* 4, nr. 10 (October 1952); Ernst Bäumler, *Ein Jahrhundert Chemie* (Düsseldorf: Econ Verlag, 1963); Gottfried Plumpe, "The I.G. Farbenindustrie A.G. as a Multinational Enterprise," European University Institute, EUI Colloquium Papers: "The Early Phase of Multinational Enterprise in Germany, France, and Italy," Florence, October 17–19, 1984.

49. From chart in George Brearley, "Preliminary Draft Report of the UK Member of the Tripartite Team appointed to consider dispersal problems relating to I.G. Farbenindustrie AG," June 27, 1950, PRO FO 371/85679.

50. Below is an overview of the factories and main products of the Middle Rhine Group plants:

Farbwerke Hoechst (main factory):	organic and inorganic chemicals nitrogen fertilizers dyes and textile intermediates pharmaceuticals pesticides solvents plastics
Mainkur	dyes and textile intermediates organic chemicals synthetic resins

Griesheim (Chem. Fabrik Griesheim)	organic chemicals and intermediates
	carbide electrodes
Offenbach (Napthol-Chemie)	dyes and organic intermediates
Griesheim (Griesheim Autogen)	apparatus for oxoacetylene welding oxygen and other technical gases
Marburg (Behringwerke)	serums and vaccines

(Table from Hermann Gross, *Material zur Aufteilung der I.G. Farbenindustrie Aktiengesellschaft* [Kiel: Institut für Weltwirtschaft, 1950], Liste VI: Heutige I.G.-Betriebe in der Bundersrepublik Deutschland.)

51. Herbert Moulton, "Subject: The Höchst Plant of I.G. Farbenindustrie A.G., 30 July 1945," pp. 6–15, Hoechst Firmenarchiv, Frankfurt-Hoechst. I was unfortunately not permitted to visit the archive in person. My thanks to the Director, Dr. Manfred Simon, for sending me a complete copy of this document.

52. On the lack of high-quality managers at the Hoechst group, ibid., pp. 24 ff. On investment, see above (chap. 1).

53. Bäumler, *Ein Jahrhundert*, pp. 134–136.

54. Kurt Lanz, *Weltreisender in Chemie* (Düsseldorf: Econ Verlag, 1978), pp. 499–500; Karl Winnacker, *Nie den Mut verlieren: Erinnerungen an Schicksaljahre der deutschen Chemie* (Düsseldorf: Econ Verlag, 1971), pp. 103–104.

55. Heinz L. Krekeler (German General Counsel, New York) to Dr. Reuter (I.G. Farben Liquidation Committee), March 31, 1951, HStA Hessen, Wiesbaden, 502/229.

56. Winnacker, *Nie den Mut*, p. 79.

57. Lanz, *Weltreisender*, p. 499.

58. Report on Kalle & Co., A.G., Wiesbaden-Biebrich, June 11, 1945, Bayerwerksarchiv (hereafter BWA), Leverkusen: Produktionsbeginn: Unterlagen.

59. PCA SHAEF, Industrial Investigation Report 16: I.G. Farbenindustrie A.G., Werk Hoechst, 1945, p. 2, in HStA Hessen, Wiesbaden, 649, OMGH Bipartite Liaison Div. ED. 17/164–1/12.

60. Herbert Moulton, "Subject: The Höchst Plant of I.G. Farbenindustrie A.G., 30 July 1945," p. 2, Hoechst Firmenarchiv, Frankfurt-Hoechst. Note that Moulton claimed that the Hoechst main plant "cannot function, on *any* scale, without the continued flow of raw materials and supplies from outside sources (French, British, or Russian zones or imports)." Ibid., p. 47.

61. "U.S. Military Government, General Order No. 2 Pursuant to Military Government Law No. 52 (Blocking and Control of Property):

I.G. Farbenindustrie, A.G.," n.d., reprinted in Hoechst AG, *Dokumente aus den Hoechst-Archiven*, vol. 48: U.S.–Administration. *Die Verwaltung des Werkes Hoechsts 1945–1953* (Frankfurt-Hoechst: Hoechst A.G., 1976), pp. 30–31. MG Law No. 52 (Amended), n.d., was simply a document that empowered the U.S. MG to seize properties in its zone. Reprinted in ibid., pp. 21–24.

62. Winnacker, *Nie den Mut*, pp. 150, 165; interview of author with Prof. Dr. Kurt Hansen, Leverkusen, December 4, 1984.

63. The trial is transcribed in large part in *Trials of the War Criminals before the Nürnberg Military Tribunals*, Vol. VII: "The I.G. Farben Case," (Washington: USGPO, 1953). See also Borkin, *Crime and Punishment*, pp. 170–198. The DuBois quote is from Borkin, p. 195. For a personal account by one of those tried (and acquitted), see Heinrich Gattineau, *Durch die Klippen des 20. Jahrhunderts* (Stuttgart: Seewald Verlag, 1983). There is further discussion of the relations between the firm and the Nazi regime in chapter 1, and of the trials themselves and their importance in chapter 6.

64. Harold Zink, *The United States in Germany, 1944–1955* (Princeton, NJ: D. Van Nostrand, 1957), pp. 24–29.

65. On the division of authority with regard to I.G. Farben in the U.S. zone, ACA, I.G. Farben Control Committee, "Minutes of the First Meeting on 4 February 1946, 25 February 1946," COIG/M(46)1, PRO FO 1005/355.

66. OMGUS ED Memo, "Subject: Division of Responsibility within the ED for Control of the Properties formerly owned by I.G. Farben, 17 December 1945," BA Koblenz, OMGUS ED Dir. Off., 3/127–3/8. U.S. I.G. Farben Control Office, "Seizure, Control and Disperal of I.G. Farbenindustrie AG in the United States Area of Occupation 5 July 1945–1 January 1950," pp. 19–20, in ibid., BICO Dec. Br. 11/11–1/12.

67. I.G. Control Office, "Seizure," pp. 19–21.

68. Ibid. See also chapter 4 on the French zone.

69. Schwarz, *Vom Reich*, pp. 149–199, for British and French postwar problems and their policy implications.

70. Ibid., pp. 39–146.

71. See, for instance, U.S. I.G. Farben Control Office, "Arbeitsanweisungen Nr. 1 [Auszugsweise Übersetzung], 4 September 1945," p. 3, in Hoechst, *Dokumente*, vol. 48, p. 66; "Property Control Officer, Hesse, to whom it may concern (appointment of German property custodian of Hoechst), 26 February 1946," in ibid., p. 81. More generally, see Zink, *United States*, pp. 26–42, esp. 31, 41–42.

72. Hessisches Wirtschaftsministerium, Abt. CIb, "Bericht für den wirtschaftspolitischen Ausschuss des Landtages betr.: I.G. Farbenindustrie A.G.," p. 1, March 26, 1947, HStA Hessen, Wiesbaden, 507/1391.

73. LWA für Gross-Hessen, "Supervision of Economics and Distribution of Goods in *Land* Greater Hesse" (translation), December 8, 1945, p. 1, HStA Hessen, Wiesbaden, 649, OMGH ED Dir. Off. 8/80-1/20.
74. See source for figure 3.
75. LWA für Gross-Hessen, "Supervision of Economics," December 8, 1945, p. 1, HStA Hessen, Wiesbaden, 649, OMGH ED Dir. Off., 8/80-1/20.
76. "Economics Offices of *Land* Greater Hesse," n.d. (ca. March 1946), HStA Hessen, Wiesbaden, 649, OMGH ED Dir. Off., 8/80-2/4.
77. Hessisches Staatsministerium für Wirtschaft und Verkehr, Geschäftsverteilungsplan, Stand 1 April 1949, HStA Hessen, Wiesbaden, 507/4159.
78. Hessisches Wirtschaftsministerium, "Aktennotiz, Betr.: Superphosphat, 28 May 1946," HStA Hessen 507/1068.
79. Dr. Wilhelmi, "Aktenvermerk, February 5, 1947," HStA Hessen, Wiesbaden, 507/2732; Abt. Industrie, "Aktenvermerk, February 20, 1947, Betr.: Belegung der Farbwerke Hoechst durch bizonale Postverwaltung," in ibid.
80. Eurgen (?) L. Weyland to Chief, Research Control Branch, "Subject: Dr. Frowein, Friedrich [Copy], 3 August 1946," HStA Hessen, Wiesbaden, 649, OMGH ED Dir. Off., 8/80-2/3.
81. Dr. Frowein to Dr. Rudolph Mueller, "Betr.: Organisation der Forschungslenkung in der amerikanischen Zone, October 8, 1946," HStA Hessen, Wiesbaden, 507/2419.
82. Coster to OMGUS ED Research Control Branch, "Subject: Beilstein Edition, 15 November 1946," HStA Hessen, Wiesbaden, 649, OMGH ED Dir. Off., 8/80-2/3.
83. Correspondence between U.S. and Hessian authorities from July to September 1947 is in HStA Hessen, Wiesbaden, 507/1798.
84. Compiled from Farbwerke Hoechst, *Annual (Research) Reports*, 1947, 1948, and 1949, HStA Hessen, Wiesbaden, 507/657[8].
85. Ibid.
86. From table in Hermann Gross, *Wirtschaftswichtige Forschung und Wissenschaftsfinanzierung in Deutschland und die USA* (Kiel: Institut für Weltwirtschaft, 1955), p. 33.

3. BUSINESS AS USUAL: THE BRITISH AND THE I.G. NIEDERRHEINGRUPPE

1. For a description of "the world in the summer of 1945" and Britain's place in it, see Alan Bullock, *Ernest Bevin: Foreign Secretary, 1945-1951* (London: Heinemann, 1983), pp. 3-19, 25-48.

2. The British zone of occupation has, until recently, received remarkably little attention from scholars. Certainly there was and remains no equivalent to John Gimbel's study of *The American Occupation of Germany* (Standford: Stanford University Press, 1968). In recent years a spate of articles and monographs has appeared and more are promised. Two of the most comprehensive collections are Claus Scharf and Hans-Jürgen Schröder, eds., *Die Deutschlandpolitik Grossbritannien und die britische Zone 1945–1949* (Wiesbaden: Franz Steiner Verlag, 1979); and Dietmar Petzina and Walter Euchner, eds., *Wirtschaftspolitik im britischen Besatzungsgebiet 1945–1949* (Düsseldorf: Schwann, 1984). The former has an excellent bibliography, albeit already a bit dated.

There is a very good dissertation on the subject of the chemical industry in the British zone, 1945–1949, by Frederick Stratmann, "Chemische Industrie unter Zwang? Staatliche Industriepolitik am Beispiel der chemischen Industrie Deutschlands 1933–1949" (Ph.D. dissertation, University of Göttingen, 1984) (now published under the same title [Stuttgart: Steiner, 1985]). A summary of some of his findings is in F. Stratmann, "Strukturen der Bewirtschaftung in der Nachkriegszeit. Das Beispiel der Chemiebewirtschaftung in der britischen und der Bizone 1945 bis 1948," pp. 153–171 in the Petzina and Euchner collection. Stratmann's work, although thoroughly grounded in historical sources, is also oriented to questions in the discipline of political science relating to corporatism and interest-group politics on the German domestic scene. It is thus less concerned than we are here with the extra-German constraints/pressures upon initial British policy and with comparisons with the other occupation zones.

3. EIPS, "Membership of EIPS," EIPS/P(44)38, November 1, 1944, Public Record Office, London (KEW) (hereafter PRO), T236/110; EIPS, "Progress Report on the Work of EIPS," n.d. (ca. June 1944), PRO BT 11/2394.

4. Foreign Office to various government departments, "Subject: EIPS, 20 March 1944," PRO T 236/110.

5. EIPS, "Report of the Working Party on the German Chemical Industry, Draft, June 1944," PRO BT 11/2392; ibid, approved version, EIPS/P(44)20, July 20, 1944, PRO FO 1005/959.

6. Record of meeting held on July 15, 1944, to discuss the setting up of the CCES, PRO BT 11/2394; CCG(BE) Trade and Industry Division, "Industry in the British Zone of Occupation, First Period June 1945–December 1946," March 1, 1947, PRO FO 1005/1884.

7. Correspondence between R. G. Somervell (undersecretary, BOT) and Sir Percy Mills, April and May 1945, PRO BT 11/2394. On the general question of industrial connections, especially for the U.S. zone, see James S. Martin, *All Honorable Men* (Boston: Little, Brown, 1950).

8. Some material on this subject is in PRO BT 11/2394.

9. On the special status of the I.G. controller in the British zone, Stratmann, "Chemische Industrie," p. 219; and ACA, I.G. Farben Control Committee, "Minutes of the first meeting, 25 February 1946," p. 1, PRO FO 1005/355.

10. Stratmann, "Chemische Industrie," p. 222; figures are compiled from various documents in file PRO FO 936/58. The French estimate is in J. P. Fouchier, "Notes pour de Chef de la Division de la Production Industrielle concernant la situation des personnels anglais et américains, 20.VI.49," French Military Government Documents, Colmar, C98, Caisse 1838, Dossier 36. There are some clues about industry connections of lower-ranking employees of the British MG. H.C. Wrather, for instance, was the team leader of the NRW Chemical Section HQ and of I.G. Farben Control for most of the occupation. He is reported to have accompanied ICI personnel to the Bayer Works in Leverkusen in 1950. E. L. Douglas Fowles to Brig. C. C. Oxborrow, "Subject: Activity of Dr. Hermann Bücher, 24 April 1950," PRO FO 371/85677.

11. *Wirtschaftsstatistik der deutschen Besatzungszonen 1945–1948* (Oberursal: Europa-Archiv, 1948), p. 45.

12. ACC CORC, "Quadripartite Control of Plants and Assets of I.G. Farbenindustrie A.G.," Memo by U.S. member, PRO FO 1039/773. It should be noted, however, that these estimates are somewhat problematic. See chapter 2.

13. On the development of the chemical industry in general in the nineteenth and early twentieth century, see the classic studies by L. F. Haber, *The Chemical Industry during the Nineteenth Century* (Oxford: Clarendon Press, 1958); and *The Chemical Industry 1900–1930: International Growth and Technological Change* (Oxford: Clarendon Press, 1971). For Bayer, see Bayer A.G., *Beiträge zur hundertjährigen Firmengeschichte 1863–1963* (Leverkusen: Bayer A.G., 1963/1964); and Farbenfabrik Bayer, "Chemiewerk aus einheitlicher Konzeption," in the special issue devoted "Zum 75.-Jährigen Bestehen des Chemie-Verbandes," *Chemische Industrie* 4, nr. 10 (October 1952): A621–A624.

14. See George Meyer-Thurow, "The Industrialization of Invention: A Case Study from the German Chemical Industry," *Isis* 73 (1982): 363–381.

15. For more on Duisberg's career, see chapter 1, and the thus-far hagiographical biographies in Hans-Joachim Flechtner's *Carl Duisberg: Vom Chemiker zum Wirtschaftsführer* (Düsseldorf: Econ Verlag, 1959); entry on Duisberg in *Neue Deutsche Biographie*, vol. 4 (Berlin: Duncker und Humblot, 1959): 181–182; FF Bayer, "Chemiewerk"; and H.-J. Flechtner's "Die Elberfeld Fabriken," p. 19, in Bayer A.G., *Beiträge*.

16. Bayer Group, "FARDIP Questionnaire," 1949, p. 15, PRO FO 1013/1557. For the integration of the Bayer production units in general, I.G. Farbenindustrie A.G. (in Auflösung), "Bericht zum RM-Schlußbilanz zum 20. Juni 1948 und DM-Eröffungsbilanz zum 21. Juni 1948," Veröffentlichung der Tripartite I.G. Farben Control Group, n.d., pp. 7–9.

17. George Brearley, "Preliminary Draft Report of the UK Member of the Tripartite Investigation Team appointed to consider Dispersal Problems relating to I.G. Farbenindustrie A.G., 26 June 1950," chart B, in PRO FO 371/85679. The Lower Rhine Group had a single factory outside the British zone, in Parchwitz in the Russian zone.

18. See chapter 1 above.

19. Michael Pohlenz, "Leverkusen und das Bayer-Werk in den Jahren 1944–1946" (M.A. thesis, University of Cologne, 1981), pp. 10–14. The Elberfeld plant of the Bayer group was not damaged by the bombing, and in fact the only damage was suffered in April 1945 when ten hand grenades were set off in the plant in skirmishes with the U.S. army. Thus, mostly windows and roofs were damaged, which were repaired for the most part by June 1945. I.G. Farben Elberfeld, "Bericht über den Stand der Fabrikation im Werk Elberfeld der I.G. Farbenindustrie A.G.," June 8, 1945, p. 1, Bayerwerksarchiv, Leverkusen (hereafter BWA), Produktionsbeginn, Unterlagen 1945.

20. Interview of author with Prof. Dr. Kurt Hansen, Bayer A.G., Leverkusen, West Germany, December 4, 1984; Pohlenz, "Leverkusen," pp. 14–17 and 24–26; "Entwurf zur Niederschrift der TDC in Leverkusen am 15 April 1945, 16 April 1945," BWA 12/13, 1: DC und TDC Protokolle.

21. Hansen interview.

22. Entwurf zur Niederschrift (see note 20).

23. On the schemes themselves, see the account by Albert Speer, *Inside the Third Reich* (New York: MacMillan, 1970; Avon Books ed., 1971), esp. pp. 560–594.

24. Hansen interview. In this portrayal of cooperation and unity among management and workers, it is, of course, possible that Hansen, a later chairman of the managing board of Bayer A.G., casts the events immediately following the war in an overly optimistic light. But the behavior of workers during negotiations with the Allies (see chapter 7) lends credence to his descriptions.

25. Power was handed over in the course of a meeting on June 12, 1945. Dr. Wingler, "Notiz über eine Besprechung im Rathaus Wiesdorf am 12. Juni 1945," BWA 206/8: Besatzung: Besuchsberichte.

26. CIOS Report, "Miscellaneous Chemicals: I.G. Farbenindustrie A.G. Elberfeld and Leverkusen Germany," April 27, 1945, by F. Curtis

and M. Fogler, May 29, 1945, Hauptstaatsarchiv [hereafter HStA] Hessen, Wiesbaden, 649, OMGH Bipartite Liaison Div., Econ. Br., 17/164–2/7.

27. I.G. Farben (Bayer) to WVChemInd Bezirksgruppe Nordrheinprovinz, "Betr.: Ihr Rundschreiben vom Juli 1946," August 7, 1946, p. 7, BWA 62/15: WVChemInd. F. Stratmann considers such percentages of plant destruction of little value (Stratmann, "Chemische Industrie," pp. 150 ff.). But in this case: (1) the percentages can give a rough idea of the percentage of destruction in relation to other plants; (2) 15 percent seems to be a fairly good, reliable figure, since it is confirmed by independent expert authorities (i.e., Bayer and Broadbent) and is consistent with the findings of Curtis and Fogler.

28. Keith Robbins, *The Eclipse of a Great Power: Britain, 1870–1975* (London: Longman, 1983), pp. 203–207 and 213–217; Bullock, *Bevin, 1945–1951*, pp. 49–80.

29. Wolfgang Rudzio, "Die ausgebliebene Sozialisierung an Rhein und Ruhr. Zur Sozialisierungspolitik von Labour-Regierung SPD 1945–1948," *Archiv für Sozialgeschichte* 18 (1978): 1–39. References to most of the relevant secondary literature are in this article. For a differing interpretation, see Horst Lademacher, "Die britische Sozialisierungspolitik im Rhein-Ruhr Raum 1945–1948," pp. 51–92, in Scharf and Schröder, eds., *Deutschlandpolitik Grossbritanniens*.

30. See Bullock, *Bevin, 1945–1951*, pp. 69–70, 74–80, 96 ff. See also R. S. Barker, "Civil Service Attitudes and the Economic Planning of the Attlee Government," *Journal of Contemporary History* 21 (July 1986): 473–486. Barker concentrates on domestic economic planning and argues that "the civil service, by its reluctance to reappraise its traditional assumptions and methods of operation and to adjust them in light of the new role demanded of it by peace-time economic management, contributed significantly to the initial weakness of the planning process" (p. 473). He stresses institutional inertia and limited creativity rather than deliberate attempts to sabotage new policies. The influence of established elite groups on the development of British policy in Germany itself would be an interesting subject for a detailed study.

31. Of course, one would not expect to find evidence of this sort surviving in the official files of the British military government.

32. BOT, "Memorandum of information affecting the control of German industry to various trade associations," n.d. (ca. May 1945), PRO BT 11/2578.

33. Technical documents collected by the teams were often microfilmed and disseminated to industry. There is various material in ibid., as well as documents on the dissemination of the information in PRO BT 211/11, 13, 17, 18, and 21. One of three cover organizations, CIOS,

BIOS, or FIAT, published summary reports of the investigations and made them available to the public. An index of the reports, published in 1946, is in PRO FO 1005/1602. The full names of the organizations mentioned are: Combined Intelligence Objectives Subcommittee; British Intelligence Objectives Subcommittee; and Field Intelligence (later Information) Agency, Technical. For the interest of the large British chemical firms in the products of three organizations, there is correspondence from 1947 on the FIAT "Reviews of German Science" in PRO FO 371/65237.

34. The file on the discontinuation of the investigations is in PRO FO 371/65032.

35. Quote from letter of Cecil Weir to J. H. Wright, July 3, 1948. This and other related correspondence is in PRO FO 371/71188. The ABCM also asked for information from the MG on its competitors. The correspondence and minutes are in BT 211/268.

36. Interview with Dr. Heinz Nedelmann, Essen, West Germany, July 9, 1984. German historian Josef Foschepoth criticizes this as one of the "persistently hawked [*nachhaltig kolportierten*] legends of the Geman postward period." Foschepoth, "Konflikte in der Reparationspolitik der Alliierten," in Foschepoth, ed., *Kalter Krieg und deutsche Frage* (Göttingen: Vandenhoeck & Ruprecht, 1985), p. 195, quoted in Volker Berghahn, *Unternehmer und Politik in der Bundesrepublik* (Frankfurt: Suhrkamp, 1985), p. 80.

37. Alan Milward contends that "There are several indications that serious competition from German exports was not expected before the middle of the fifties," in Milward, "Grossbritannien, Deutschland und der Wiederaufbau Westeuropas," in Petzina and Euchner, eds., *Wirtschaftspolitik*, p. 29. His comment does not hold for the chemical industry.

38. W. A. M. Edwards (ICI) to COFGA, "Subject: Manufacture of HCH in German and ICI Patent Rights, 11 February 1947," PRO FO 943/132.

39. Notes of a meeting held at the Board of Trade to discuss German dyestuffs, February 22, 1946, PRO FO 942/454.

40. S. Courtauld to Stafford Cripps (BOT), February 13, 1946, PRO BT 211/90.

41. A. J. C. Walters to Sir F. H. E. Woods (BOT) and enclosure, March 28, 1946, in ibid.

42. Further material for 1946 is in PRO FO 942/498; for 1947, in PRO BT 211/268.

43. S. Cripps to S. Courtauld, March 5, 1946, PRO BT 211/90.

44. Eric Seal to W. Ritchie (COFGA), July 24, 1946, PRO FO 943/135.

45. Minutes of meeting, July 3, 1946, PRO BT 211/90.

46. COFGA, Interdepartmental Working Party on Economics, "Minutes of Meeting on 30 April 1946," EWP/M(46)8, May 3, 1946, pp. 3–4, PRO FO 942/498.

47. Report of visit of Rayon Federation, August 27–September 6, 1947, PRO BT 211/329; Report of the Dyestuffs Mission to Germany, March 6–18, 1948, PRO BT 211/268. Note, however, that the decision to allow both groups to go was taken by September 1946: J. Davidson Pratt (ABCM) to Derek Wood (BOT), November 4, 1947, in ibid.

48. Derek Wood to R. Keeling, November 7, 1947, PRO BT 211/268.

49. This is an aspect of the British occupation comparatively well-treated in the secondary literature. Thus, discussion here is brief. Bullock, *Bevin, 1945–1951*; Petzina and Euchner, introduction in Petzina and Euchner, eds., *Wirtschaftspolitik*, pp. 7–10; Hans-Peter Schwarz, *Vom Reich zur Bundesrepublik*, 2d enl. ed. (Stuttgart: Klett-Cotta, 1980), pp. 149–175; Josef Foschepoth, ed., *Kalter Krieg und deutsche Frage. Deutschland im Widerstreit der Mächte 1945–1952* (Göttingen: Vandenhoeck & Ruprecht, 1985), esp. the essays in Part I.

50. The connection among the three is dealt with to some extent in R. W. Carden, "Before Bizonia: Britain's Economic Dilemma in Germany, 1945–1946," *Journal of Contemporary History* 14 (July 1979): 535–555.

51. Bullock, *Bevin, 1945–1951*, p. 150.

52. The order for control is reprinted in Pohlenz, "Leverkusen," Anlage 9. Also MG Germany, British Zone of Control, I.G. Farbenindustrie A.G., "Directive to Control Officer, 22 November 1945," PRO FO 1039/644. For the U.S. influence, note the comment of Treasury attaching "a scheme for the taking control of I.G. Farben in a particular way. The scheme has been forced on us by the Americans." Minutes of September 14, 1945, PRO T 236/974.

53. On the breakup of the I.G., see chapter 7. The British attitude is expressed, for instance, in C. G. McFarlane (FO German Commercial Relations and Industry Dept.), "Minutes on Meeting with Mr. Brearley, 7 June 1950," PRO FO 371/85678.

54. Nedelmann and Hansen interviews. Karl Winnacker, *Nie den Mut verlieren: Erinnerungen an Schicksaljahre der deutschen Chemie* (Düsseldorf/Wien: Econ Verlag, 1971), p. 150.

55. There is no full-length biography of Haberland available. The following is compiled on the basis of information in Haberland's resume from September 1960 in BWA 271/2: Personalia, Dr. Haberland; and the entry on him in *Neue deutsche Biographie*, vol. 7 (Berlin: Duncker und Humblot, 1965): 392–393.

56. Hansen interview. On the more general problem of postwar planning by German industry (and government), see Ludolf Herbst, *Der totale*

Krieg und die Ordnung der Wirtschaft (Stuttgart: Deutsche Verlags-Anstalt, 1982).

57. See chapters 5, 6, and 7.

58. Bayer, A.G., *Geschichte des Werkes Uerdingen der Farbenfabriken Bayer Aktiengesellschaft* (Uerdingen, 1956), pp. 113-114.

59. Winnacker, *Nie den Mut,* pp. 117-118, for Winnacker's relationship with and evaluation of Haberland; p. 165 for his job at Knapsack.

60. Hansen interview.

61. Sir Gordon Macready to high commissioner, April 5, 1951, PRO FO 1036/180. It should be noted that, although all of the members of the managing board of the I.G. were tried at Nuremberg, other I.G. employees who were not board members were also tried. One example is Heinrich Gattineau.

4. THE TECHNICAL LIMITS TO EXPLOITATION

1. When one considers furthermore the relative lack of interest by French historians in their own zone, and the fact that the French were the last of the former western Allies to open their archives to researchers, the dearth of scholarly literature on the zone becomes comprehensible. The remarkably meager French language literature on the subject was for the most part written in the occupation period itself or shortly thereafter (e.g., Alfred Grosser, *L'Allemagne de l'Occident 1945-1952* [Paris: Gallimard, 1953]; André Piettre, *L'économie allemande contemporaine—Allemagne occidentale—1945-1952* [Paris: Génin, 1952]); or else is part of an autobiography or memoir of one of the contemporary actors (e.g., Hervé Alphand, *L'étonnement d'être. Journal 1939-1973* [Paris: Librairie Fayard, 1977]). An exception is the more recent, but still rather dated work by Adalbert Korff, *Le revirement de la politique française a l'égard de l'Allemagne entre 1945 et 1950* (Ambilly-Annemasse: Imprimerie Franco-Suisse, 1965). Korff's book is still very useful to the understanding of the postwar rapprochement of the two powers. Marc Hillel, in *L'occupation française de L'Allemagne* (Paris: Balland, 1983), has tried to interest the French again in the period. The book is, however, of little use for scholars: it is for the most part composed of anecdotes combined with heavy reliance on the literature from the 1950s, and especially upon Grosser. It is furthermore limited in availability in Germany and appears to have been available only in limited numbers even in France.

The most important book on the occupation remains an early work by an American historian, F. Roy Willis, *The French in Germany, 1945-1949* (Stanford: Stanford University Press, 1962). The book's long-standing reputation is well deserved. The situation has begun to change somewhat. See the papers in Claus Scharf and Hans-Jürgen Schröder, eds., *Die Deutschlandpolitik Frankreichs und die französischen Zone 1945-1949* (Wiesbaden: Steiner, 1983).

On the opening of French archives, see Raymond Poidevin, "Sources relatives a l'étude de la France devant l'Europe sous la IVe République," in Walter Lipgens, ed., *Sources for the History of European Integration (1945–1955)* (Leyden: Sijthoff, 1980), pp. 29–30. The French archives for the period were opened at least in theory after a change in the French archive law in 1979. Most of U.S. materials had long been available, while British sources had opened gradually beginning in 1975. In practice, the French records have become generally available only since about 1982.

2. Alan Milward, *The Reconstruction of Western Europe, 1945–1951* (London: Methuen, 1984), pp. 126–167, esp. p. 167.

3. I do not deal here with the Saarland, which the French treated as a part of their own national economy in the early years of the occupation. For industry in general (and the chemical industry in particular) in South Baden during the occupation, see Rudolph Laufer's detailed study, *Industrie und Energiewirtschaft im Land Baden 1945–1952: Südbaden unter französischer Besatzung* (Freiburg/München: Verlag Karl Alber, 1979). The major contours of his argument are summarized in Laufer, "Die südbadische Industrie unter französischer Besatzung 1945–1949," pp. 141–153, in Scharf and Schröder, eds., *Deutschlandpolitik*.

4. French Military Government, Sous-direction Chimie, "Note sur les industries chimiques en Zone Française d'Occupation," November 4, 1948, Archives of the French Foreign Ministry, Archives de la Haut Commissariat de la Republique Française en Allemagne, Division Production Industrielle de la Direction Général des Affairs Economiques et Financières, Colmar (hereafter MGFr), Caisse 1838, Dossier 29 (hereafter 1838/29): Index Z.F.O. 1948/49.

5. Commandement en Chef Français en Allemagne, Office Militaire de Securité, Element Français, "Etude d'ensemble sur la potential industriel de guerre de la Zone Française d'Occupation," July 1949, MGFr 1538/3: Potential industriel de guerre, 1945/1948.

6. Chef de la sous-Division des Industries Chimiques to M. le Directeur de la Prod. Ind., "Objet: Représentation allemande de la ZFO auprès de la Bizone, 16 August 1948," MGFr 2066/4: Chimie, Chrono.: 1.VII.–31.VIII.1948.

7. This and the following are drawn from a number of sources, including BASF A.G., "BASF schreibt Geschichte," pp. A577–A582 of the special issue devoted "Zum 75.-Jährigen Bestehen des Chemie-Verbandes," *Chemische Industrie* 4 (October 1952); and Joseph Borkin, *The Crime and*

Punishment of I.G. Farben (New York: Free Press, 1978; Pocket Books ed., 1979).

8. United States Strategic Bombing Survey, USSBS Rpt. 117: *Ludwigshafen-Oppau Works of I.G. Farbenindustrie AG, Ludwigshafen, Germany*, 2d ed. (Washington, DC, January 1947), p. 4.

9. For synthetic fuels development and the I.G., see Peter Hayes, *Industry and Ideology: I.G. Farben in Nazi Era* (New York: Cambridge University Press, 1987), pp. 36–42, 115–119, 133–145, and 183–185; and Thomas P. Hughes, "Technological Momentum in History: Hydrogenation in Germany, 1898–1933," *Past and Present* 44 (August 1969): 106–132. For synthetic rubber development, see the dissertation by Peter Morris, "The Development of Acetylene Chemistry and Synthetic Rubber by I.G. Farbenindustrie Aktiengesellschaft, 1926–1945" (dissertation, Oxford University, 1982). For this, and also for a critique of Hughes's argument, see Gottfried Plumpe, "Industrie, technischer Fortschritt und Staat: Die Kautschuksynthese in Deutschland 1906–1944/45," *Geschichte und Gesellschaft* 9 (1983): 564–597.

10. The subsidy turned out to be even larger than BASF or the government had first believed: BASF and the Little I.G. paid back some government loans to build plants (e.g., those that had financed the construction of the massive Leuna-Werke) in near-worthless currency during the inflation following the war. See Peter Hayes, *Industry and Ideology: I.G. Farben in the Nazi Era* (New York: Cambridge University Press, 1987), pp. 10–13.

11. Hughes, "Technological Momentum;" Borkin, *Crime and Punishment*; Helmuth Tammen, *Die I.G. Farbenindustrie AG 1925–1933* (Berlin: Verlag Helmuth Tammen, 1978), pp. 139–144.

12. Tammen, ibid.; Dieter Petzina, *Autarkiepolitik im Dritten Reich: Der nationalsozialistische Vierjahresplan* (Stuttgart: Deutsche Verlags-Anstalt, 1968).

13. "BASF schreibt Geschichte," p. A580.

14. "Rapport Production Industrielle," July 29, 1946, p. 3, MGFr 530/ Doc. 2: Rapports.

15. This and the above comparison are drawn from F. Chardin, "La place de la BASF dans l'industrie chimique," January 10, 1951, MGFr 1840/46: Place de la BASF dans l'industrie chimique.

16. Mission Diplomatique Française auprès des Gouvernements Alliés, "Le Problem Allemand," August 21, 1944, p. 2, in Archives of the French Foreign Ministry, Paris, Série Y, 1944–1949, Dossier 278 (hereafter FFM Y/278): Politique des alliés 2.–12.1944.

17. For initial French policy, see Schwarz, *Vom Reich*, pp. 179–199, esp. p. 179; Willis, *The French*.

18. Mission Diplomatique Française auprès des Gouvernements

Alliés, "Le Problem Allemand," August 21, 1944, p. 2, FFM Y/278.
19. "Etude de la politique à suivre vis-à-vis des industries chimiques allemandes," n.d. (mid-1945), MGFr 2049/14: Divers, chimie.
20. Ibid.
21. Ibid.
22. Ibid.
23. This feeling continued well into 1947 at least. "Compte-Rendu du Congrés I.G.-Chimie," April 19, 1947, MGFr 1840/45: Conseil de Surveillance I.G. 1948.
24. The ambivalence of French policy is now beginning to be explored in the literature, although in areas other than industrial policy. See the contribution by Alain Lattard, "Gewerkschaften und Betriebsräte in Rheinland-Pfalz 1945-1947. Zur französischen Gewerkschaftspolitik in Deutschland nach dem Zweiten Weltkrieg," in Scharf and Schröder, eds., *Deutschlandpolitik*, pp. 155-184. Rainer Hudemann also deals in a general way with the ambiguities of French policy—and briefly with the ambiguities of French economic policies—in his article, "La zone française d'occupation sous le premier gouvernement du Général de Gaulle (mai 1945 à janvier 1946)," *Etudes gaulliennes* 6 (1978): 25-37.
25. Destruction to Hoechst and Leverkusen, the central factories of the other two former I.G. plant complexes in western Germany, was surprisingly slight. See above, chapters 2 and 3. Leuna, the largest plant of the former I.G. in the Soviet zone, was also less heavily damaged than the BASF factories in the French zone. A French investigatory team in mid-1946 estimated the overall damage at ca. 10 percent. Scheffer-Rapport No. 12, "Compte-Rendu du voyage à Leuna, 8-10 Juillet 1946," MGFr 1840/27: Industrie de l'azote en Allemagne, 1946/47.
26. See Werner Abelshauser, *Wirtschaftsgeschichte der Bundesrepublik Deutschland 1945-1980* (Frankfurt: Suhrkamp, 1983), p. 38, for general chemical industry figures. For BASF figures, see the sources listed in table 5.
27. Dr. Kaumann (Abt. Interzonenhandel) to Dr. Maltzan, "Betr.: Demontage von Oppau, December 15, 1947, Bundesarchiv (BA) Koblenz, Z8/1964, Bl. 89.
28. Abelshauser, "Wirtschaft und Besatzungspolitik," p. 116, in Scharf and Schröder, eds, *Deutschlandpolitik*.
29. Summary tables are in MGFr 1840/33-34: BASF Etat exportations et ventes 1947-1948; for the French-French-zone trade, see Friedrich Jerchow, *Deutschland in der Weltwirtschaft 1944-1947. Alliierte Deutschland- und Reparationspolitik und die Anfänge der westdeutschen Aussenwirtschaft* (Düsseldorf: Droste Verlag, 1978), pp. 427-441.
30. For the French side of the affair, the documents in MGFr 1840/39: Livraisons d'ammoniaque 1947-1948. For the U.S. side, OMGUS ED

Dir Off./EA, 3/145–2/20.

31. Conseil de Surveillance de l'I.G. Farben, "Questions commerciales et financières," 20 May 1948, MGFr 1840/45: Conseil de Surveillance I.G. 1948; for the program for recruiting German technicians and scientists to work in France, the monthly reports of the Division Prod. Ind. for 1946 in MGFr 530/2: Rapports. Ironically, Francolor was a dye trust that the I.G. had imposed upon French industry in 1941. See Hayes, *Industry and Ideology*, pp. 279–285.

32. J. Gandilhon, "Rapport sur les conditions générales de fonctionnement de l'Administration Française de la BASF," October 13, 1949, MGFr 1840/63: Rapport 1949.

33. The contents of the file in MGFr 2871/1: Direction des Industries Chimiques 1945/1953.

34. Compiled on the basis of information in MGFr 1838/1: Compétence S/Div. Chimie et autres sections 1946/47.

35. MGFr Commission, "Réorganisation du contrôle de l'économie allemande, Procès-Verbal de la lle Scéance, Extrait," February 17, 1947, MGFr 2871/7: Contrôle des industries Allemandes (1947/52).

36. Koenig to M. le Delegué Général pour le GM de l'Etat Rhéno-Palatin, "Objet: Répartition de produits contingentés à BASF," February 6, 1948, MGFr 2066/1: Chimie 2.1.–29.2.1948.

37. MGFr 1840/41–43: BASF: Rapports d'activité 1946–1948.

38. S/Dir des Ind. Chim., "Compte-Rendu de visite au Bipartite Control Office," June 6, 1948, MGFr 1838/36: Production Bizone 1948–49; J. P. Fouchier, "Note pour le Chef de la Div. de la Prod. Ind., Reseignements concernant la situation des personnels anglais et américains," June 20, 1949, in ibid.; Fouchier, "Evolution du contrôle de l'industrie chimique en Allemagne," June 2, 1949, MGFr 2871/7: Contrôle des industries allemandes 1947/1952.

39. This argument is based on admittedly incomplete information; but there are some personnel files indicating this in MGFr 1838/2: Office Militaire de Sécurité. Généralités 1948–1952.

40. Material in file MGFr 2066/1: Chimie 2.1.–28.2.1948. The file is representative of long series of very thick folders on the day-to-day activities of chemical control personnel.

41. The BASF archivists contacted in 1983/1984 claimed that they have no holdings for the immediate postwar period; the French apparently seized them. The records do not, though, appear in the French military government archives.

I have been able to find evidence only of token resistance to French control, and that only very late in the occupation. BASF officials showed reluctance by 1949 to send reports to Baden-Baden and sent them (if they sent them at all) in German rather than in (occasionally very bad) French. Their behavior is, though, perfectly understandable. The entire file, in MGFr 1840/44: BASF Rapport d'activité 1949, is worthy of note in this context, and especially Haut-Comm. Fr., Dir. Général des Affaires Economiques et Financières to M. le Dir. Gén. de la BASF à Ludwigshafen, "Objet: Relations entre la BASF et les autorités administratives," August 24, 1949, in ibid.

My thanks in particular to Professor Richard Hamilton for suggesting that I pursue this line of inquiry.

42. W. von Haken, "Die Katastrophe von Ludwigshafen," *Angew. Chem.*, B20 (September 1948): 244–245.

43. This explains why the figures for production of primary nitrogen reported in table 5 do not reflect the impact of the explosion: nitrogen production centered on Oppau, as did a significant part of the firm's methanol production.

44. "Bulletin Mensuel du Commandement en Chef Français en Allemagne," vols. 1–4 (1945–1949), passim.

45. Usine d'Oppau, "Rapport du Directeur Technique pour le mois de Déc. 1946," December 20, 1946, MGFr 1840/41: BASF, rapport d'activité 1946.

46. Service Entretien, "Rapport du mois de janvier 1948," February 9, 1948, MGFr 1840/43: BASF, Rapport d'activité 1948.

47. S/Direction des Services Techniques, "Rapport du mois de Mai 1948," June 11, 1948, in ibid.

48. S/Dir. des Services Techniques, "Rapport du moi de Juillet 1948," in ibid.

49. Summary tables in MGFr 1840/33–34: BASF Etat exportations et ventes 1947–1948.

50. For information on Krekeler, see Wolfgang Mommsen, *Die Nachlässe in den deutschen Archiven*, Teil II (Boppard: Harald Boldt Verlag, 1983), pp. 905–906; Memo from Randolph Newman (U.S.-zone I.G. Farben controller) to Kelleher (U.S. chief of decartelization section), "Subject: Dr. Heinz Krekeler, 28 June 1950," BAKoblenz, OMGUS Bico Dec. 11/11–2/8; for correspondence, MGFr 539/C45–46: Relations commerciales ZFO/Bizone.

51. For example, Chef de la S/Div des Industries Chimiques to M. le Directeur de la Prod. Ind., "Objet: Représentation allemande de la ZFO auprès de la Bizone," August 16, 1948, MGFr 2066/4: Chimie, Chrono. 2.7.–31.8.1948.

52. F. Roy Willis, *France, Germany and the New Europe 1945–1967*, rev. and exp. ed. (Stanford: Standford University Press, 1968), p. 19.

53. Schwarz, *Vom Reich*, pp. 179–199.

Notes to pages 109–119   241

5. THE CONTINUING CRISIS

1. The increased influence of the United States on bizonal policy was especially evident after the conclusion in May 1947 of the second bizonal agreement (which increased the American contribution to the bizonal budget). See Falk Pingel, "Der aufhaltsame Aufschwung," in Dietmar Petzina and Walter Euchner, eds., *Wirtschaftspolitik im britischen Besatzungsgebiet 1945–1949* (Düsseldorf: Schwann, 1984), pp. 57–59.

2. There are many problems with this index. For a discussion of some of them, see chapter 6.

3. Werner Abelshauser, *Wirtschaftsgeschichte der Bundesrepublik Deutschland 1945–1980* (Frankfurt: Suhrkamp, 1983), pp. 32–45.

4. This problem is examined in more detail in chapter 6.

5. LWA Hessen, "22. Wochenbericht (2.6–7.6.1947)," October 10, 1947, pp. 7–8, Hauptstaatsarchiv (hereafter HStA) Hessen, Wiesbaden, 507/141.

6. LWA Hessen, "2. Monatsbericht," February 3, 1948, p. 2, HStA Hessen, Wiesbaden, 507/143.

7. Henning Dieter, "Problems in Connection with Coal Distribution," p. 6, n.d. (ca August 1947), HStA Hessen, Wiesbaden, 649, OMGH ED IndBr., 8/94–1/16.

8. The statistics are fairly complete for the Hoechst factory from mid-1945 to mid-1948, although there is no breakdown of production levels for particular products and only by *product line*.

9. "Übersicht über die Produktion der Farbwerke Hoechst" (ca. July 1948), Hoechst Firmenarchiv, Frankfurt-Hoechst, *ohne Signatur* (o.S.).

10. Ibid.

11. Bayerwerk Leverkusen, "A-Fabrik," 1945/1946; 1947; 1948/49, Bayerwerksarchiv (hereafter BWA), Leverkusen, o.S. There was a similar development in the production formaldehyde, which is produced on the basis of methanol, and which was restricted by the Allies.

12. This argument is developed in more detail at the end of this chapter.

13. ACC Law No. 9 "Providing for the Seizure of Property Owned by I.G. Farbenindustrie and the Control thereof," October 30, 1945, in Hoechst A.G., *Dokumente aus Hoechst-Archiven*, vol. 48: "US–Administration. Die Verwaltung des Werkes Hoechst 1945–1953" (Frankfurt-Hoechst: Hoechst, 1976), p. 73.

14. The Allied Control Authority, Decartelization Commission, I.G. Farben Control Committee (ACA DECO COIG) Minutes 1946–1948 are in Public Record Office, London (Kew) (hereafter PRO) FO 1005/355–360.

15. ACA DECO COIG, "Minutes of 3d Meeting, 29–30 April 1946," May 15, 1946, COIG/M(46)3, pp. 5–6, PRO FO 1005/355.

16. ACA DECO COIG to DECO, "Report for the Foreign Ministers' Conference, 3 Feb. 1947," COIG/P(47)1, (British original, French publication), pp. 1–2, PRO FO 1005/356.

17. ACA DECO COIG, "Minutes of 8th Meeting, 29–30 July 1946," August 5, 1946, COIG/M(46)8, pp. 8–9, PRO FO 1005/355.

18. See Friedrich Stratmann, "Chemische Industrie unter Zwang? Staatliche Industriepolitik am Beispiel der chemischen Industrie Deutschlands 1933–1949" (dissertation, University of Göttingen, 1984)—now published under the same title with minor alterations (Stuttgart: Steiner, 1985), pp. 231–232.

19. BICO joint secretary to chairman, Verwaltungsrat, "Betr. Errichtung einer Zweimächte-Organisation," August 5, 1948, in Hoechst A.G., *Dokumente*, vol. 49: "Der Hoechst Konzern entsteht, Teil I" (Frankfurt-Hoechst: Hoechst, 1978), pp. 29 ff. Note that the creation of FARDIP occurred at the same time, and that both were a specific response to the breakdown of four-power cooperation.

20. See the last part of this chapter.

21. The function, personnel, and structure of these organizations have already been described in some detail by Friedrich Stratmann in "Chemische Industrie," pp. 241–252, 284–287.

22. Ibid.

23. The protocol of the first LFA meeting is in VfW, Abt. Chemie, "Bericht über die Sitzung des FA bzw. LFA Chemie, Hattenheim, 8.VIII.47," Bundesarchiv (BA) Koblenz, Z8/531, Bl. 21–26.

24. Stratmann, "Chemische Industrie," p. 293.

25. Moldenhauer's diary, available in the Bundesarchiv, covers the postwar period (and Moldenhauer's role in control of the I.G.) in volume 4, *Nachlass Moldenhauer/4*, BA Koblenz. Moldenhauer had been a director of the I.G. and served as Reich finance minister in 1929/1930. For some further information on his career, the other volumes in his *Nachlass* (1–3); and Peter Hayes, *Industry and Ideology: I.G. Farben in the Nazi Era* (Cambridge/New York: Cambridge University Press, 1987), pp. 54, 61–62, and 97–98. My thanks to Dr. Martin Vogt for drawing my attention to these records.

26. This is described in more detail in chapter 6.

27. "Kurzbericht über die ausserordentliche Sitzung des LA Chemie am 4.III.1948," p. 2, BA Koblenz Z8/618, Bl. 49.

28. LWA Hessen, "7. Wochenbericht (9.2.–14.2.48), 17.II.1948," HStA Hessen 507/143.

29. LWA Hessen, "Halbmonatsbericht (23.2.–28.2.48), 2.III.48," p.

15, HStA Hessen 507/143.

30. Kurzbericht über die ausserordentliche Sitzung des LA Chemie am 4.III.1948." p. 2, BA Koblenz Z8/618, Bl. 49.

31. Friedrich Stratmann and his coauthors use this concept in their article on the British zone Sparta-Plan of 1946. See Alexander Drexler, Wolfgang Krumbein, and Friedrich Stratmann, "Die britischen 'Sparta-Pläne' 1946," pp. 245–263 in Josef Foschepoth and Rolf Steininger, eds., *Die britische Deutschland- und Besatzungspolitik 1945–1949* (Paderborn: Schöningh, 1985).

32. Nicholas Balabkins, *Germany under Direct Controls: Economic Aspects of Industrial Disarmament, 1945–1948* (New Brunswick, NJ: Rutgers University Press, 1964), pp. 112 ff.; a more recent description and analysis of the postwar coal situation is in Werner Abelshauser, *Der Ruhrkohlenbergbau seit 1945* (Munich: Beck, 1984), pp. 15–49.

33. Henry Wallich, *Mainsprings of the German Revival* (New Haven: Yale University Press, 1955; Greenwood Press Reprint, 1976), p. 367.

34. VAW Abt. Chemie, "Tätigkeitsbericht August 1947," September 6, 1947, BA Koblenz Z8/495, Bl. 23.

35. "Bericht über die Sitzung des FA bzw. LWA Chemie am 8.VIII.47," pp. 3–4, BA Koblenz Z8/531, Bl. 22.

36. BECG (signed Brig. J. C. Cowley, Chair) to Chairman, Executive Committee for Economics, "Betr.: Volle Ausnutzung knapper Reserven in den Vereinigten Zone," May 20, 1947, p. 1 (Abschrift), BA Koblenz Z8/531, Bl. 27.

37. Ibid.

38. WVChem Ind (BK), "Geschäftsbericht für die Zeit Juli bis Dez. 1947," p. 16, Archiv des Verbandes der Chemischen Industrie (ArchVChemInd), Frankfurt.

39. "Note on the Second Meeting of the Working Party set up to report on the allocations, usage, and requirements of coal to the chemical industry in the Combined Zone," October 17, 1947, BA Koblenz B103/VI/5/48.

40. I.G. Leverkusen (Dr. Loehr) an das VAW Abt. Statistik u. Planung z.Hd. von Herrn Dr. Keiser, "Betr.: Kohlenversorgung des Werkes Leverkusen," April 2, 1947, BA Koblenz B103/VI/5/59, Bd. 3.

41. Dr. Hauck, VAW Abt. Chemie, "Aktennotiz über Besprechung am 9.IV.47 mit Herrn Dipl.-Ing. List von I.G. Farbenindustrie A.G. Leverkusen," April 10, 1947, BA Koblenz B103/VI/5/57, Bd. 2.

42. I.G. Farben Leverkusen, Ing.-Abt. K (signed Warnecke and ?) to VAW Ref. Chemie z. Hd. von Herrn Dr. Hauck, "Betr.: Kohlenbedarfszahlen Werk Leverkusen," May 2, 1947, BA Koblenz B103/VI/5/59, Bd. 3.

43. *Konzept* of letter from VAW Hauptabt. Aussen- und Interzonenhandel (Dr. von Maltzan) to I.G. Leverkusen, June 16, 1947, BA Koblenz Z8/1964, Bl. 114.

44. Dr. Hauck (VAW, Abt. Chemie), "Aktenvermerk über Besprechung am 27.VI.47, Betr.: Kohlenzuteilung Leverkusens," BA Koblenz B103/VI/5/57, Bd. 2.

45. Ibid.

46. I.G. Leverkusen, "Direktionskonferenz minutes," October 10, 1947; "Technische Direktionskonferenz minutes" (from October 21, November 11, and December 9, 1947), BWA Leverkusen 12/13, 5.

47. Länderrat der amerik. Zone, Hauptausschuss Wirtschaft, Arbeitsausschuss Kohle und Bergbau, Unterausschuss Kohle, "Meeting on 5.XII.47," BA Koblenz Z1/615, Bl. 2–10.

48. Ibid.

49. "Protokolle über die Sitzung des Arbeitsstabs Kohle, Mannheim, am 6.IV.48," BA Koblenz Z1/615, Bl. 112–113.

50. Statistics compiled from "Mitteilungsblatt des VAW/der VfW des VWG," issues from May 24, 1947, to August 30, 1948, inclusive, in BA Koblenz Z8/1321, Bl. 34–330, passim. Allocations to the chemical industry do not include those to its synthetic fibres or fertilizer segments. Furthermore, the allocations are "planned" ones, that is not actual deliveries. Finally, the number of tons is not in hard coal units (HCU), but rather is the gross number of tons of coal regardless of type.

6. CONSOLIDATION OF RECOVERY

1. 714 HQ Land NRW, Chemical Industries Section, "Notes on a meeting with Field Team Officers, *Land* NRW, 5.II.1948," Public Record Office, London (Kew) (hereafter PRO) FO 1013/1261.

2. Norman A. Shepard, "German Chemical Industry: Capacity High, Recovery Slow," *Chemical Industries* (January 1948), p. 52 (in Bayerwerksarchiv (BWA), Leverkusen, O. Bayer/13).

3. "German Chemical Importance Dims," *The Journal of Commerce* (New York), November 12, 1948, in ibid.

4. See note 1.

5. See note 2.

6. See note 3.

7. Karl Häuser in Gustav Stolper, Karl Häuser, and Knut Borchardt, *The German Economy, 1870 to the present* (New York: Harcourt, Brace, and World, 1967), pp. 219–228.

8. Ibid., pp. 211–212.

9. Ibid., p. 213.

10. Knut Borchardt, in Stolper and others, *The German-Economy*, p. 236, states the idea of the conjunction of these three factors explicitly, although Borchardt is more cautious than Häuser in claiming that they were the cause of the economic miracle.

11. Matthias Manz, "Stagnation und Aufschwung in der franzoesischen Besatzungszone von 1945 bis 1948" (dissertation, University of Mannheim, 1968)—published under the same title, edited by Werner Abelshauser, ed. (Ostfildern: Scripta Mercaturae Verlag, 1984), pp. i–iii, 114–122. Werner Abelshauser, *Wirtschaft in Westdeutschland 1945–1948. Rekonstruktion und Wachstumsbedingungen in der amerikanischen und britischen Zone* (Stuttgart: Deutsche Verlags-Anstalt, 1975). Abelshauser's arguments on the misleading nature of the statistics at around the time of the currency reform were anticipated in contemporary reports: LWA für Hessen, Abt. Berichtswesen und Statistik an alle Empfänger des Halbmonatsberichts des LWA, "Betr.: Industrieproduktion im Monat Juli," August 23, 1948, Hauptstaatsarchiv (hereafter HStA) Hessen, Wiesbaden, 507/143.

12. Albrecht Ritschl, "Die Währungsreform von 1948 und der Wiederaufstieg der westdeutschen Wirtschaft. Zu den Thesen von Matthias Manz und Werner Abelshauser über die Produktionswirkungen der Währungsreform," *Vierteljahrshefte für Zeitgeschichte* (January 1985): 136–165; see also Werner Abelshauser's response, "Schopenhauers Gesetz und die Währungsreform," pp. 214–218 in ibid.

13. There is a technical description of the construction of the industrial index in Research and Statistics Section, Industry Branch, Office of the Economic Advisor, Frankfurt, BAOR 21, "The Index of Industrial Production for the Federal Republic of Germany," June 1950, in French Military Government Archives, Colmar (seen at the Foreign Ministry Archives in Paris), Caisse 1838, Dossier 30: Index Trizone, 1949/1950. The report notes the weaknesses of the index. For instance, "The [index] weights reflect the ratios between the respective total net values added of industries in 1936 (the net value added being the value or work done in an industry, excluding intake from other industries)." The weakness is that the ratios had changed in the postwar period from those of 1936, and the use of contemporary ratios would produce a different index. The assumption was that proportions of net output to gross output remained the same as in 1936, something that could not be verified. It would be very helpful to have a scholarly examination of this problem of index-construction and weighting and its effect on interpretations of the economic development of western Germany in the postwar period.

14. Little work has been done on this problem, although some is beginning to appear. See for instance Wolfgang Benz, "Zwangswirtschaft und

Industrie: Das Problem der Kompensationsgeschäft am Beispiel des Kasseler Spinnfaser-Prozesses von 1947," *VfZ* 32 (July 1984): 422–440. For a first-hand account (and indeed this is the basis for Benz's article), see Ludwig Vaubel, *Zusammenbruch und Wiederaufbau. Ein Tagebuch aus der Wirtschaft* (Munich: R. Oldenbourg, 1984), esp. pp. 111 ff.

15. Such a conclusion is not, of course proven by this analysis of the chemical industry. Studies of other industrial sectors would clarify if the chemical industry is or is not a special case.

16. "Niederschrift über die 15. Verkaufsbeschprechung in Leverkusen am 5. März 1948," BWA 12/15, Nr. 1–21.

17. "Niederschrift über die 17. Verkaufsbesprechung in Leverkusen am 4. Juni 1948," in ibid.

18. "Niederschrift über die 18. Verkaufsbesprechung in Leverkusen am 2. Juli 1948," in ibid.

19. Ibid.

20. "Niederschrift über die 19. Verkaufsbesprechung in Leverkusen am 26. Juli 1948"; "Niederschrift über die 20. Verkaufsbesprechung in Leverkusen am 17. September 1948"; "Niederschrift über die 21. Verkaufsbesprechung in Leverkusen am 20. Oktober 1948," in ibid.

21. Farbwerke Hoechst, "Survey of developments in the year 1948 at Farbwerke Hoechst," enclosure to letter from M. Erlenbach to OMGH Chemical Section (Dr. Scholz), January 28, 1949, HStA Hessen, Wiesbaden, 649, OMGH EDIndBr 17/25–2/7.

22. Werner Abelshauser, *Wirtschaftsgeschichte der Bundesrepublik Deutschland* (Frankfurt: Suhrkamp, 1983), pp. 55–56. Abelshauser's figures do not agree fully with those presented for the ERP in table 11, but Abelshauser includes both ERP and Mutual Security Agency (MSA) aid to December 31, 1952 in his table. The MSA replaced the ECA at the end of 1951. See W. A. Brown, Jr., and Redvers Opie, *American Foreign Assistance* (Washington, DC: The Brookings Institution, 1953), pp. 505–539.

23. For more on this, see Henry Wallich, *Mainsprings of the German Revival* (New Haven: Yale University Press, 1955; Greenwood Press reprint, 1976). Wallich is also one of the few early writers to stress this important aspect of the Marshall Plan aid. See pp. 364–366.

24. "Stand der Investitionskredite," *Die chemische Industrie* Nr. 5 (November 1949), no page numbers, BWA, Leverkusen, 15/9.3: Finanzwesen, Korrespondenz, etc. 1949–1952.

25. Josiah E. DuBois, Jr., *Generals in Grey Suits* (London: The Bodley Head, 1953), pp. 345–346; S. Balke, "Der I.G.-Farben-Prozess in Nürnberg," *Chemie-Ingenieur-Technik* 21 (January 1949): 34; Borkin, *Crime and Punishment*, pp. 194–195. Borkin inexplicably does not include one of the guilty I.G. executives in his list of sentences and those found guilty. See

DuBois or Balke for a more accurate list. See chapter 1 for a discussion of the relations between the I.G. and the Nazis, and chapter 2 for a discussion of the place of the Nuremberg trials in American occupation policy.

26. Borkin, pp. 176, 187–188. The court was not unified in its verdict. One judge dissented from the others in their findings on the slave labor charge and claimed in his opinion that all members of the I.G. managing board should have been found guilty. Ibid., pp. 196–198.

27. Ibid., p. 177.

28. Ibid, p. 178. "Auschwitz" came third among the five counts. Borkin quotes Emanuel Minskoff of the prosecution staff in a conversation with DuBois.

29. DuBois was a lawyer from Camden, New Jersey, who was asked by Bernard Bernstein, the assistant general counsel at the Treasury Department, to help the Department out early in 1941. He stayed for six years as chief counsel of Foreign Funds Control, where he was responsible for the seizure of I.G. assets in the western hemisphere. DuBois, *Generals*, pp. 14–15.

30. Ibid., pp. 33 and 35. Underlined portion—his emphasis.

31. Ibid.

32. "Freigesprochener wieder Direktor," *Die Neue Zeitung* (München), Nr. 78 (September 21, 1948), Institut für Weltwirtschaft, Kiel, Zeitungsausschittssammlung, Key word: I.G. Farbenindustrie A.G., Mappe VI, 1947–1949.

33. A list of the wartime and postwar positions of all of the defendants is in Paul Danek, "Zur reaktionären Rolle des staatsmonopolistischen Kapitalismus bei der Wiedererrichtung und Machtausweitung des IG-Farbenmonopols in Westdeutschland" (Dissertation, University of Halle, 1961), Anlage II.

34. The case of Max Ilgner is interesting in this context. Ilgner, a former I.G. managing board member and previous head of Farben's Berlin NW 7 office (which controlled the firm's intelligence, propaganda, and political economic operations), had been sentenced at Nuremberg to three years' imprisonment for his role in plundering the European chemical industry. By mid-1949, he was once again free (as was his uncle, former I.G. managing board chairman Hermann Schmitz) but never again participated directly in the chemical industry. Instead, he ran the planning office of the *Evangelischen Hilfswerk*, a Protestant organization that promoted relief and recovery in postwar Germany. Ilgner counted among the members of his planning committee Dr. Ulrich Haberland, Bayer's director, and Fritz Gajewski, former head of Farben's Sparte III who had been tried and found innocent on all counts at Nuremberg. Gajewski himself had a number of positions in the chemical industry after the war. He was

chairman of the managing board at Chemieverwaltungs-A.G., Frankfurt, and was member of the supervisory board of a number of companies, among them the former I.G. subsidiary Chemische Werke Hüls A.G. Information on Ilgner came from correspondence (personal and confidential letter of May 5, 1949, from Ilgner to Richard Merton and enclosure, a *Rundschreiben* to the members of the planning committee from May 6, 1949 [sic]) in the Historisches Archiv of the Metallgesellschaft A.G., Frankfurt, Korrespondenz Richard Merton, I-L, 1.I.1949–Juni 1950.

35. In this, the industry was not that unusual. See Volker Berghahn's excellent analysis of West German industrial elites in general in *The Americanisation of German Industry, 1945–1973* (Cambridge/New York: Cambridge University Press), pp. 40–71. He also discusses the problem of contact and influence on the part of those closed out from direct participation in postwar business affairs in ibid., esp. pp. 53–62.

36. In this very important way, the German occupation differed from the occupation of Japan. In the latter case, top officials were removed from large firms, never to return. Regardless, the question of the extent to which Japanese society changed after the war is also open. My thanks to Professors Mansel Blackford and James Bartholomew for bringing this to my attention.

37. Bipartite Control Office, Joint Secretariat, an den Vorsitzer des Verwaltungsrates für den Präsidenten des Wirtschaftsrates, "Betr.: Errichtung einer Zweimächte-Organization und eines Zweizonen-Ausschusses für die Kontrolle des I.G.-Farben-Besitzes im Vereinigten Wirtschaftsgebiet," August 5, 1948, Hoechst A.G., *Dokumente aus Hoechst-Archiven* vol. 49: "Der Hoechst Konzern entsteht, Teil 1" (Frankfurt-Hoechst: Hoechst, 1978), pp. 29–31.

38. Ibid.

39. Myron Maupin (U.S. I.G. Farben control officer) to Mr. Richardson Bronson, chief, U.S. Decartelization Element, "Subject: Background of FARDIP Personnel," July 4, 1949, BA Koblenz OMGUS BICO Dec 11/11–1/5.

40. Ibid.

41. Ibid.

42. Ibid.

43. Loehr played a prominent role in various top-level meetings at the Leverkusen plant throughout the occupation period.

44. Correspondence in BA Koblenz, B 102/337, especially the letter from Dr. Prentzel to P. Dencker of September 3, 1948.

45. See W. O. Reichelt, *Das Erbe der I.G. Farben* (Düsseldorf: Econ, 1956), pp. 79 ff. The advisers included Brecht and Bücher from FARDIP, as well as H. J. Abs (member of the managing board of the Deutsche

Bank, formerly of the I.G. supervisory board, and later of the BASF supervisory board), and W.A. Menne (head of the chemical industry trade association and later member of the managing board of Hoechst A.G.).

46. Myron Maupin to Richardson Bronson, "Subject: Unofficial Meeting with the French Control Officer," April 9, 1949, BA Koblenz OMGUS BICO Dec, 11/11–2/4.

47. See chapter 4.

48. A simple set of statistics will clarify this: in 1950, when West German industry in general was producing at 140 percent of its 1936 rate, the coal industry had just reached its 1936 rate. Wallich, *Mainsprings*, p. 367. For a more detailed picture of the coal situation, see Balabkins, *Germany under Direct Controls* (New Brunswick, NJ: Rutgers, 1964); and, more recently, Werner Abelshauser, *Der Ruhrkohlenbergbau seit 1945* (Munich: Beck, 1984).

49. Werner Abelshauser, "Korea, die Ruhr und Erhards Marktwirtschaft. Die Energiekrise von 1950/1951," *Rheinische Vierteljahrsblätter* 45 (1981): 287–316.

7. NEW BEGINNINGS AND RESURGENCE

1. Dr. Ulrich Haberland, "Westdeutschlands chemische Industrie um die Jahrewende," *Chemie-Ingenieur-Technik* 23 (1951): 23–24.

2. *Chemiewirtschaft in Zahlen* (Zusammengestellt vom Verband der chemischen Industrie, e.V.), 4th ed. (Düsseldorf: Econ, 1960), pp. 38–39. Note: for the table from which this information was garnered, 1938 = 100.

3. Discrepancies between the indices for the chemical industry in the Federal Republic in tables 15 and 16 are accounted for by the fact that the figures in table 15 are the result of internal statistical collection and manipulation in the federal government, while the figures in table 16 are OEEC figures.

4. John Gimbel, *The American Occupation of Germany* (Stanford: Stanford University Press, 1968). See also, chapter 2.

5. J. P. Fouchier, "Evolution du Contrôle de l'industrie chimique en Allemagne," June 2, 1949, French Military Government Records, Colmar (seen at the Quai d'Orsay in Paris), C98, Caisse 2871, Dossier 7: Contrôle des industries allemandes 1947/1952.

6. This point is made in the dissertation from Freiburg (1981) published a year later by Georg Mueller, *Die Grundlegung der westdeutschen Wirtschaftsordnung im Frankfurter Wirtschaftsrat 1947–1949* (Frankfurt: Haag und Herchen, 1982).

7. The Military Security Board was set up according to a directive

of January 17, 1949, by the military governors of the three western zones of occupation. It was directed against threats to security from within Germany. "Threats to security" were construed broadly and included prevention of the resurgence of paramilitary organizations, of militarily useful industrial production, and of research with military applications. The MSB had three divisions: Military, Industrial, and Scientific Research. See the "Directive by the Governors of the Three Western Zones on the Organization of the Allied Military Security Board," January 17, 1949, in Beate Ruhm von Oppen, ed., *Documents on Germany under Occupation 1945–1954* (London: Oxford University Press, 1955), pp. 350–355. For more information, see "Stellungnahme zur Errichtung eines alliierten militärischen Sicherheitsamtes in Deutschland," part of the "Protokoll of the Ministerpräsidentenkonferenz in Hamburg 11–12 February 1949," in *Akten zur Vorgeschichte der Bundesrepublik Deutschland*, Vol. V (Munich: Oldenbourg, 1981), pp. 180–185.

The negotiations surrounding the creation of the MSB for the chemical industry would be a useful topic for scholarly inquiry. The French, it is clear, wanted very much a control group for chemicals on a par with those for iron and steel and for coal in the High Commission period (French note to the British and Americans, "Observations Françaises au Sujet du contrôle de l'industrie chimique allemande," June 14, 1949). The British, in a note to the French Foreign Ministry from their embassy in Paris, stated that they believed that "adequate machinery is already available for achieving allied objectives in this field," and suggested that such a move would be detrimental to the new German state (note of July 11, 1949). The Americans insisted that the security aspects of the chemical industry were handled adequately in the MSB, and stressed that the pursuit of decartelization efforts within the industry would be more effective in curbing the power of the chemical industry than a control group. They also criticized the French for not implementing decartelization measures to an adequate extent in their zone (Henri Bonnet, U.S. ambassador to France, to Robert Schuman, aide memoir, July 8, 1949). The French seem to have capitulated in this area rather suddenly and completely, although the reason is not clear. In fact, J. P. Fouchier, the head of the chemical industry subdivision within the division Production Industrielle of the French military government, complained in late August 1949 to his chief about plans to trim control personnel for the chemical industry (not including the personnel administering the I.G. firms) to just five men by October, down from fifty in late 1948/early 1949 (Fouchier to Directeur, August 26, 1949). All citations from French Military Government Records, Colmar (seen in Paris), C98, Caisse 2871, Dossier 7: Contrôle des industries allemands 1947/1952.

8. For the text of the agreement as promulgated by the Allied Military Governors in Germany on April 13, 1949, see U.S. Department of State, Office of Public Affairs, *Germany 1947–1949: The Story in Documents* (Washington: USGPO, 1950), pp. 366–371.

9. Dr. H. Nedelmann, "Verbote und Demontagen im Bereich der Kohlechemie," *Chemie-Ingenieur-Technik* 21 (August 1949): 317–319.

10. Once again, this is an area about which we know little. There is a considerable body of material on the subject in the Bayerwerksarchiv, Leverkusen, in the records of the Fachverband Kohlechemie.

11. For a discussion of the IAR and the diplomacy surrounding it, see for instance Raymond Stokes, "German Energy in the Postwar U.S. Economic Order, 1945–1951," *Journal of European Economic History*, forthcoming.

12. "Abschrift der Niederschrift der Abmachungen zwischen den Alliierten Hohen Kommissaren und dem Deutschen Bundeskanzler auf dem Petersberg am 22. November 1949," BWA, Leverkusen, Records of the Fachverband Kohlechemie, 186/K1.19.2.

13. On the ECSC, see Volker Berghahn, *The Americanisation of West German Industry, 1945–1973* (Cambridge/New York: Cambridge University Press, 1987), pp. 119–154.

14. For the chemical industry and the federal government in this regard see "Chemische Industrie protestiert gegen Produktionsbeschraenkungen,". Bundeswirtschaftsministerium, *Tages-Nachrichten* (February 2, 1951), in BA Koblenz, B102/3838.

15. I.G. Chemie, Papier, und Keramik, Hrsg. *Jahrbuch 1951*, p. 14. (I.G. Chemie-Bibliothek, Hannover).

16. The following section summarizes some of the themes developed and presented piecemeal in the preceding chapters. It also relies heavily on Hans-Dieter Kreikamp, "Die Entflechtung der I.G. Farbenindustrie A.G. und die Gründung der Nachfolgegesellschaften," *Vierteljahrshefte für Zeitgeschichte* 25 (April 1977): 220–251.

17. *Geschichte des VEB Leuna Werke "Walter Ulbricht" 1945 bis 1981* (Leipzig: VEB Deutscher Verlag für Grundstoffindustrie, 1986), p. 21.

18. Richardson Bronson to control officer, "Subject: Request by Mr. Martin for consideration of dispersal by industries," January 21, 1946, reprinted in *Dokumente aus Hoechst-Archiven*, vol. 49: "Der Hoechst Konzern entsteht, Teil 1" (Frankfurt-Hoechst: Hoechst, 1978), pp. 19–20.

19. Hans Kreikamp recognizes the importance of the developments of 1948 but downplays them somewhat since, he claims, there was no real change in Allied policy to break up the I.G. into a large number of small companies or in the fact that the Germans had only limited influence on the proceedings. Kreikamp, "Die Entflechtung," pp. 223 f.

20. Ibid., pp. 225–227.

21. Interview of author with Dr. Felix Prentzel, July 17, 1986, Frankfurt a.M.

22. Prentzel played an important role in deciding the membership of FARDIP, for instance. Correspondence and memoranda on this subject from 1948 are in BA Koblenz, B 102/337.

23. Prentzel, "Vermerk für den Herrn Staatssekretär (Schlafejew?), (Vertraulich)," January 18, 1950, BA Koblenz B 102/338.

24. Prentzel interview.

25. Kreikamp, "Die Entflechtung," pp. 234–235, 248–249. The trade unions representing chemical workers tended to support industrialists and representatives of the federal government, especially in the idea of *Kerngesellschaften*. See Kreikamp, p. 234. Prentzel interview.

26. A text of the report, as well as a fairly complete set of commentaries on it from various German experts and organizations, is available in the Handakten Prentzel (DL3), Degussa Firmenarchiv, Frankfurt.

27. Kreikamp, "Die Entflechtung," p. 229.

28. See note 25; Kreikamp, pp. 229–231.

29. Kreikamp, "Die Entflechtung," pp. 220–251.

30. See, e.g., ibid., pp. 220, 242.

31. Ibid., pp. 249–250.

32. Ibid., pp. 223, 233–234, 248–249.

33. See, e.g., ibid., pp. 221, 247.

34. Lothar Franke, "Reorganisation der US-Dekartellisierungabteilung" (newspaper clipping, source unidentified), February 6, 1950, Degussa-Archiv, Frankfurt, 151/014. See also Volker Berghahn, *Americanisation*, p. 165.

35. On the ambivalence during the New Deal period, for instance, see Ellis Hawley, *The New Deal and the Problem of Monopoly* (Princeton: Princeton University Press, 1966).

36. Note, for example, the outcome of the prosecution of the Standard Oil trust in 1911.

37. Kreikamp, p. 230. Interestingly enough, there is evidence that at least some of those on the Allied side who had to do with the *Entflechtung* came to believe that the setting up of three separate *Kerngesellschaften* of approximately equal size would in fact held *preserve* the breakup of the I.G. As McFarlane noted in his memo of the conversation with George rearley on June 6, 1950, "It is the opinion of Mr. Brearley that the Frankfurt plants [i.e., the Maingau Works Group] should be combined in one complex under strong direction which is entirely lacking at present. It is his belief that any deconcentration or leaving of the other two complexes alone without a combination of the Frankfurt plants would create the very

opportunity of the construction of I.G. Farben under another name so soon as the allied control disappears. On the other hand, if a strong complex in the Frankfurt area could act as a counterpoise to the other two complexes then the eventual chances of cartelization are much diminished." McFarlane memo, June 7, 1950, PRO FO 371/85678. The same argumentation is used in the experts' report: "It is considered that the development of the Main Group to create an organization of sufficient strength as to insure its continued existence and progress side by side with the two strong organizations of Farbenfabrik Bayer and BASF is of fundamental importance to the deconcentration and dispersal program." "Report of the tripartite investigation team appointed to consider dispersal problems relating to I.G. Farbenindustrie A.G. in accordance with the terms of reference TRIFCO/MEMO(50)4 of 24 April 1950," August 1950, page 7, in French MG Records, Colmar, C96, Caisse 221, Dossier: Rapport des experts alliés.

38. On the founding of the Hoechst group, see the collection of documents assembled by the firm's archive and published as Hoechst A.G., "Der Hoechst-Konzern entsteht. Die Verhandlungen über die Auflösung von IG-Farben und die Gründung der Farbwerke Hoechst AG 1945–1953," 2 parts, in *Dokumente aus Hoechst-Archiven*, vols. 49 and 50. The Maingruppe, and especially the importance of Knapsack to it, is discussed in more detail in chapter 2.

39. The inclusion of Dormagen was, however, by no means self-evident to the Allies. Only after significant protest from the federal government did they agree to its inclusion in the new Bayer firm. See, e.g., Kreikamp, "Entflechtung," pp. 244–245.

40. Ibid., see note on p. 246.

41. On the technical details of the refounding of the companies, "Schema der Gründung der neuen BASF," n.d. (ca. 1950), in French MG Archives, Colmar, C96, Caisse 221, Dossier: Rapports des experts alliés.

42. Karl Winnacker, *Nie den Mut verlieren: Erinnerungen an Schicksaljahre der deutschen Chemie* (Düsseldorf/Vienna: Econ-Verlag, 1971), pp. 213–216.

43. Ibid. Winnacker's account of the meeting in his memoirs seems to be accurate based on contemporary documentation. See Winnacker's "Aktennotiz über Besprechung mit Liquidatoren und Aktionär-Beirat am 14.XII.1952 in Frankfurt," pp. 219–220 of Hoechst, *Dokumente*, vol. 50.

44. See entries on Hoechst A.G. and BASF in *Handbuch der Grossunternehmen*, 4th ed., "Band: BRD and West Berlin" (Darmstadt: Hoppenstedt, 1955).

45. Das Spezial-Archiv der deutschen Wirtschaft, *I.G. Farbenindustrie Aktiengesellschaft. Abwicklungsbericht und Nachfolgegesellschaften* (Heppenheim [Bergstrasse]: Das Spezial-Archiv, 1952), no page number. Seen in

Rheinisch-Westfälisches Wirtschafts-Archiv, Köln, Geschäftsbericht, box 35.

46. Ibid.

47. Ibid.

48. Interview of author with Professor Dr. Kurt Hansen at the Bayerwerk, Leverkusen, December 4, 1984.

49. See Wolfgang Heintzeler, *Der rote Faden. Fünf Jahrzehnte: Staatsdienst, Wehrmacht, chemische Industrie, Nürnberg, Marktwirtschaft, Mitbestimmung, Kirche* (Stuttgart: Seewald, 1983).

50. Peter J. T. Morris, "The Development of Acetylene Chemistry and Synthetic Rubber by I.G. Farbenindustrie Aktiengesellschaft, 1926–1945," (Dissertation, Oxford University, 1982), pp. 111–117, 125–126 (*re* Reppe's work in Ludwigshafen).

51. Kurt Lanz, *Weltreisender in Chemie* (Düsseldorf/Vienna: Econ-Verlag, 1978), passim. (See p. 52 for Sammet.)

52. Some of the implications of worldwide growth are explored in chapter 8.

53. Figure for 1943 from Hermann Gross, *Material zur Aufteilung der I.G. Farbenindustrie Aktiengesellschaft* (Kiel: Institut für Weltwirtschft, 1950), Tabelle Ia; that for 1953 compiled from Alfons Metzner, *Die chemische Industrie der Welt*, Vol. I: *Europa* (Düsseldorf: Econ, 1955), pp. 139, 147, 165. Kurt Pritzkoleit makes a similar comparison in *Bosse, Banken, Börsen: Herren über Geld und Wirtschaft* (Vienna: Kurt Desch, 1954), p. 27. Pritzkoleit's turnover figure for I.G. in 1943 is far too high, however.

54. Figures compiled from BWM IV B 1, "Aktenvermerk: Investitionen für die chemische Industrie," March 1, 1953, BA Koblenz, B102/9561, Heft 1.

55. K. H. Forster, *Finanzierung durch Abschreibungen: Nach den Ergebnissen von DM-Bilanzen* (Stuttgart: C. E. Poeschel-Verlag, 1953). Werner Abelshauser stresses the importance of the self-financing of industry (in part through tax write-offs) as well in his *Wirtschaftsgeschichte der Bundesrepublik Deutschland* (Frankfurt: Suhrkamp, 1983), pp. 72–74.

56. Forster, p. 2.

57. Ibid., p. 13.

58. "57. Direktorialsitzung in Frankfurt," May 18, 1949, pp. 448–449, fn. 5, in *Akten zur Vorgeschichte*, Vol. V.

59. Forster, *Finanzierungen*, p. 71.

60. There is the added problem of fluctuation in the market value of the stock, something I address only in passing here.

61. Peter Hayes, *Industry and Ideology: I.G. Farben in the Nazi Era* (Cambridge/New York: Cambridge University Press, 1987), p. 17; Hel-

muth Tammen, *Die I.G. Farbenindustrie AG 1925-1933* (Berlin: Verlag H. Tammen, 1978), pp. 17-19.

62. Helge Pross, *Manager und Aktionäre in Deutschland: Untersuchungen zum Verhältnis von Eigentum und Verfügungsmacht* (Frankfurt: Europäische Verlagsanstalt, 1965), p. 199, n. 36.

63. Gross, *Material*, Tabelle Ia. Even in the war years, the company's dividends were relatively high, amounting to 6-8 percent per year. For more on dividend policy, see chapter 1.

64. See "Die I.G. Farbenindustrie Aktiengesellschaft und ihre Nachfolgegesellschaften," Sonderdruck aus dem fünfbändigen *Handbuch der deutschen Aktiengesellschaften*, Jg. 1952/1953 (Darmstadt: Das Spezial-Archiv der deutschen Wirtschaft, 1953), p. 9. Seen in Rheinisch-Westfälisches Wirtschaftsarchiv, Köln, Geschäftsberichte, box 35.

65. Ibid., p. 3.

66. Ibid., p. 3.

67. Metzner, *Die chemische Industrie*, pp. 138-164.

68. Some of the best literature on the plans of the U.S. policymakers for a postwar economic order and Germany's place in it includes: John Lewis Gaddis, *The United States and the Origins of the Cold War* (New York: Columbia University Press, 1972), esp. pp. 10-31; Alfred E. Eckes, Jr., *A Search for Solvency: Bretton Woods and the International Monetary System, 1941-1971* (Austin/London: University of Texas Press, 1975); Michael J. Hogan, "American Marshall Planners and the Search for a European Neocapitalism," *American Historical Review* 90 (1985): 44-72; and Volker Berghahn, *Americanisation*. The questions of whether or not the American liberal world economic order ever really functioned, and if so for how long and how effectively, are the subject of more recent controversy. See, for instance, Alan Milward, *The Reconstruction of Western Europe, 1945-1951* (London: Methuen, 1984); and Fred Block, *The Origins of International Economic Disorder* (Berkeley, Los Angeles, and London: University of California Press, 1977). Most historians would, however, agree that American policymakers attempted to set up such a system in the immediate postwar period, and that the attempt itself must be understood to deal with major aspects of postwar economic history.

69. Olav Maseng (dep. dir. ED Hesse) to minister of Economics and Transport for *Land* Greater Hesse, "Subject: Participation of former I.G. Farben plants in Wiesbaden Exhibition," August 27, 1946, HStA Hessen, Wiesbaden, 649, OMGH ED, Office of the Director, 8/80-1/19.

70. Minute, "German trade marks" (by Mr. A. L. Burgess [BOT] to Mr. Sanders [BOT], March 6, 1946, PRO BT 11/2654).

71. OMGUS ED CO IG Farben, "Plan for the Dissolution of Bayer

Sales Organization," March 1946, p. 130, in BA Koblenz, OMGUS ED Dec. Br. 17/221-2/27.

72. Ibid.

73. "Accord on Treatment of German-Owned Patents," U.S. Department of State, *Treaties and Other International Acts Series*, no. 2415, pp. 1-3.

74. Patentanwalt Dr.-Ing. von Kreisler, "Der gewerbliche Rechtschutz in Nachkriegs-Deutschland," *Angewandte Chemie*, Teil B, 19 (January 1947): 26-27. The creation of a patent office was a matter of pressing concern for Germans both within industry and in the state apparatus in the years after 1945. See the numerous articles on the subject in *Angewandte Chemie* and *Chemie-Ingenieur-Technik*, especially in 1947 and 1948, and the discussions in the Bizonal *Exekutivrat* in Frankfurt in late 1947 and through 1948 in *Akten zur Vorgeschichte der Bundesrepublik Deutschland 1945-1949*, Bundesarchiv and Institut für Zeitgeschichte, eds., 5 vols. (Munich/Vienna: R. Oldenbourg Verlag, 1976-1983), vols. 3 and 4, passim.

75. In some areas of the world, including the United States, Bayer lost the right to use the Bayer Cross trademark. Bayer purchased the U.S. rights to use the trademark once again in early 1986.

76. Alfred von Nagel, *Indanthren. Komplexfarbstoffe. Tenside*, Schriftenreihe des BASF-Archivs, nr. 2 (Ludwigshafen, 1968), p. 17.

77. Dr. Friedrich Frowein, "Schutzrechte im Ausland," *Chemie-Ingenieur-Technik* 21 (February 1949): 80-81.

78. Material on "Gesetz über die Errichtung von Annahmestellen für Patent-, Gebrauchsmuster- und Warenzeichenanmeldungen," n.d. (ca. April 1948), BA Koblenz, Z13/87, Bd. 1, Bl. 141-149.

79. "Schlussbericht über die Tätigkeit der Annahmestelle Berlin," October 15, 1949, BA Koblenz, Z22/228.

80. "Besprechung der Militärgouverneure mit bizonalen Vertretern in Frankfurt," April 13, 1949, in *Akten zur Vorgeschichte* 5: 375.

81. Norman A. Shepard, the technical coordinator of American Cyanamide, noted the demand for German products in his article "German Chemical Industry: Capacity High, Recovery Slow," *Chemical Industries* (January 1948), p. 52, in BWA, Leverkusen, O. Bayer/13.

82. Kurt Lanz, *Weltreisender*, pp. 12-53, 67 ff., and passim.

83. Interview with Professor Dr. Kurt Hansen at the Bayerwerk, Leverkusen, December 4, 1984.

84. Metzner, *Die chemische Industrie*, pp. 162-163, 179-180.

85. "Deutsch-französisches Chemiegespräch," *Die Welt* (Hamburg), nr. 130, June 7, 1950. Seen in Zeitungsausschittsammlung des Instituts für Weltwirtschaft in Kiel, Key word: Deutschland, Westzonen, Allg. Chem. Prod. Allg., 1.5.45-31.12.51.

86. "Strukturveränderung in der Chemie-Industrie. Europäische

Zusammenarbeit befürwortet," *Der Tagesspiegel* (Berlin) nr. 1405 (August 11, 1950). Seen in ibid.

87. See contract noted in chapter 4. See also above in this chapter on Francolor's participation in the Indanthren agreement.

88. In 1948, Höchst got the license to produce penicillin from the U.S. Merck company (there is also a Firma Merck in Darmstadt—the two had ceased by then to be connected to one another)—Direktorialkanzlei des Verwaltungsrates (gez. Dr. Krautwig) to BICO Joint Sec (Betr.: BICO/Sec. [48] 574), December 3, 1948, BA Koblenz Z4/86. The moves toward getting the license started much earlier, indeed around the beginning of 1946 (see "9. Tagung des Länderrats des amer. Besatzungsgebietes in Stuttgart," June 4, 1946, *Akten zur Vorgeschichte* I: 557; James S. Martin, *All Honorable Men* [Boston: Little, Brown, 1950], pp. 210 ff.).

89. "Ein bedeutsamer Vertrag, Bayer-Schenley," *Handelsblatt* (Düsseldorf), nr. 82 (September 14, 1949). See note 83. The *Handelsblatt* hailed the contract for the fact that it represented something more than a simple agreement between two firms. "Wichtiger aber ist fast noch die psychologische Bresche, die der Vertrag in die bisherigen Isolierung der deutschen Industrie vom Ausland schlägt, nachdem zuvor das amerikanische Richtertribunal in Nürnberg die gegen die IG und damit auch gegen Bayer erhobenen Vorwürfe als unberechtigt erklärt hatte" (!). It also noted that such contracts would bring much-needed American technology to Germany and open up the U.S. market for German firms. The paper continued: "Es ist zu hoffen, dass dieser Vertrag Schule macht." One thing not mentioned is that the rights to the Bayer cross trademark in the USA remained the property of Sterling Drug Company until 1994, when Bayer reacquired it for ca. $1 billion.

90. "Start der Erdölchemie in Westdeutschland," *Frankfurter Allgemeine Zeitung* 22 (September 29, 1953), in ibid.

91. This argument is developed in more detail through comparison of the West German oil and coal industries in my article, "German Energy in the Postwar U.S. Economic Order," *Journal of European Economic History* (forthcoming).

8. FROM COLLAPSE TO COMPETITIVENESS

1. The Germans of course were not alone in this. These have been characteristics of the postwar chemical industry around the world.

2. The following is compiled from a number of sources, including BASF *Geschäftsbericht* 1985; Hoechst *Annual Report* 1985; John Davenport, "The Chemical Industry Pushes into Hostile Country," *Fortune* 19 (April

1969): 108–114, 156–162; Bayer A.G., *Bayer in Wirtschaft und Gesellschaft* (1975); Bayer A.G., *Geschichte des Werkes Uerdingen* (Uerdingen, 1956); and Alfons Metzner, *Die chemische Industrie der Welt*, Vol. I: *Europa* (Düsseldorf: Econ, 1955), pp. 160, 179.

3. Information for 1964 and 1974 from Volker Berghahn, *The Americanisation of West German Industry, 1945–1973* (Cambridge/New York: Cambridge University Press, 1986), p. 337; the figures he reports correspond closely with those I compiled from *Fortune* magazine, although he gives sales in deutsche marks rather than dollars. I compiled these and other figures from "The Fortune Directory," *Fortune* 72 (July 1965): 150; "The Fortune Directory Part II," ibid. 72 (August 1965): 170; "The 500 Largest Industrial Corporations [in the U.S.]," ibid. 91 (May 1975): 210–211; "The 300 Largest Industrial Corporations outside the U.S.," ibid. 92 (August 1975): 156; "The Fifty Largest Industrial Companies in the World," ibid.: 163; "The 500 Largest U.S. Industrial Corporations," ibid. 111 (April 29, 1985): 266–267; "The 50 Largest Industrial Corporations [in the world]," ibid. 112 (August 19, 1985): 179; "The International 500," ibid.: 183.

4. Thomas McCraw, *Prophets of Regulation* (Cambridge, MA: The Belknap Press of Harvard University Press, 1984); Ellis Hawley, *The New Deal and the Problem of Monopoly* (Princeton, NJ: Princeton University Press, 1966).

5. Brigadier C. C. Oxborrow to R. S. Swann (FO, London), "Subject: Law 35—Reorganisation of I.G. Farbenindustrie," February 27, 1953, Public Record Office (PRO), London (Kew), FO 1036/184.

6. *Basisdaten. Zahlen zur sozio-ökonomischen Entwicklung der Bundesrepublik Deutschland* (Bonn-Bad Godesberg: Verlag Neue Gesellschaft, 1974), p. 151.

7. See Gilbert Burck, "Chemicals: The Reluctant Competitors," *Fortune* 68 (November 1963), pp. 149–151.

8. For more on the U.S. chemical industry's forays into the European market, see Davenport, "The Chemical Industry Pushes into Hostile Country," pp. 108–114, 156–162.

SELECTED BIBLIOGRAPHY

PRIMARY SOURCES
Manuscript Collections

West Germany

1. Records of the Bayer A.G., Bayerwerksarchiv (BWA), Leverkusen.
    Titel 11.   Aufsichtsrat und Verwaltungsrat
    Titel 12.   Vorstand, Direktion, Direktorium
    Titel 22.   Patente—Warenzeichen
    Titel 53.   Technik—Energien
    Titel 61.   Handel und Wirtschaft, Verschiedenes
    Titel 62.   Wirtschaft-Verbände-Vereine
    Titel 64.   Wirtschaft-Reichsbehörden
    Titel 76.   Politik/Staats- und Parteipolitik, Verschiedenes
    Titel 186.   Akten des Fachverbandes Kohlechemie
    Titel 205.   Demontage, Reparationen, Besatzung Frieden
    Titel 206.   Besatzung
    Titel 271/2. Personalia: Ulrich Haberland
    Akten Professor Dr. Otto Bayer
    Akten des Sekretariats Dr. Haberland
2. Records of the Federal Government, Bundesarchiv (BAK), Koblenz.
    B102.   Bundeswirtschaftsministerium
    B103.   Bundesstelle für den Warenverkehr der gewerblichen Wirtschaft
    R7Anh.   Ministerial Collecting Center
    Z1.   Länderrat des amerikanischen Besatzungsgebietes
    Z2.   Akten des Zonenbeirates der britischen Besatzungszone
    Z3.   Akten des Wirtschaftsrates des Verwaltungsrates des Vereinigten Wirtschaftsgebietes
    Z4.   Länderrat des Vereinigten Wirtschaftsgebietes
    Z8.   Verwaltungsrat für Wirtschaft des Vereinigten Wirtschaftsgebietes
    Z10.   Zentral-Haushaltsamt für die britische Zone
    Z11.   Personalamt des Vereinigten Wirtschaftsgebietes

Z12. Büro der Ministerpräsidenten des amerikanischen, britischen, und französischen Besatzungsgebietes
Z13. Direktorialkanzlei des Vereinigten Wirtschaftsgebietes
Z14. Der Berater für den Marshallplan beim Vorsitzer des Verwaltungsrates des Vereinigten Wirtschaftsgebietes
Z19. Büro der deutschen Geschäftsstellen des Marshallplanes und des Beauftragter für den Aussenhandel der drei Länder der französischen Zone
Z22. Rechtsamt der Verwaltung des VWG
ZSg126. Presseausschnittssammlung. Institut für Weltwirtschaft der Universität Kiel
ZSg127. Presseausschnittssammlung. Verein zur Wahrung der Interessen der Chemischen Industrie Deutschlands e.V./ Wirtschaftsgruppe Chemische Industrie

Nachlass Paul Moldenhauer
Nachlass Hermann Pünder
Microfiche collection of U.S. National Archives, Record Group 260, Office of Military Government for Germany (U.S.), OMGUS [central records, bizonal administration]
All. Proz. I. Hauptabt. IX: Akten des I.G. Farben-Prozesses

3. Records of the German Trade Union Federation, Archiv des Deutschen Gewerkschaftsbundes, Düsseldorf.

I.G. Chemie, Papier, Keramik—Protokolle und Jahrbücher. 1946–1952
"Der Kontakt. Mitteilungsblatt für die Funktionäre der I.G. Chemie-Papier-Keramik." 1948–1949
Various other files

4. Records of Degussa A.G., Firmenarchiv, Frankfurt.

DL 3. Baerwind/29: Aktennotizen Baerwind, Dekartelisation
DL 3. Prentzel/1–3: Handakten und Drucksachen Dr. Prentzel
JW 46.4/2: Chemiewerk Homburg
151: Dekartellisierung
Drucksachen

5. Records located at the Friedrich Ebert Stiftung, Archiv der Sozialdemokratie, Bonn-Bad Godesberg.

Akten des Parteivorstandes der SPD: Bestände E. Ollenhauer, F. Heine, K. Schumacher

6. Records located at the Hessian Main State Archives, Hauptstaatsarchiv Hessen (HStA Hessen), Wiesbaden.

Abteilung 502. Hess. Ministerpräsident, Staatskanzlei
Abteilung 507. Wirtschaftministerium
Abteilung 649. Microfiche copy of U.S. National Archives, Record

Selected Bibliography 261

                Group 260, Office of Military Government, U.S., for Hesse (OMGH)
7. Records located at the Institut für Weltwirtschaft, Kiel.
    Zeitungsausschnittssammlung
    Demontage-Archiv
8. Records located at the Institut für Zeitgeschichte, Munich.
    Bestand MF-260. OMGUS-Akten (see BAK listing)
    Akten des Anklagers, Nürnberger Prozesse
9. Records of the German Chemical Industry Trade Union, Archiv und Bibliothek der I.G. Chemie, Papier, Keramik, Hannover.
    Geschäftsberichte. 1946–1950
    Jahrbuch. 1951
    Protokolle. 1946 and 1948.
10. Records of the Max Planck Society, Bibliothek und Archiv zur Geschichte der Max-Planck-Gesellschaft, Berlin (W).
    Excerpts from: "Die Geschichte der Kaiser-Wilhelm-Gesellschaft und Max-Planck-Gesellschaft 1945–1949." Unveröffentlichte Festgabe zum 70. Geburtstag von Otto Hahn am 8.III.1949
    A2. Akten der Generalverwaltung der MPG
    "Erinnerungsnotiz Dr. E. Repondek zur Geschichte der Kaiser-Wilhelm-Gesellschaft und der I.G. Farben, niedergeschrieben 12.XI.1970"
11. Records of Firma Merck, Firmenarchiv, Darmstadt.
    Nr. 5. Direktionskonferenz. Protokolle
    Nr.10. Wiederaufbau. Dekartellisierungsgesetz
    Schmall, W. "Meine Tätigkeit bei Merck in den Jahren 1932 bis 1965." Typescript, 1970
12. Records of Metallgesellschaft A.G., Historisches Archiv, Frankfurt.
    Korrespondenz Richard Merton, 1939–1952
    Richard Merton. Artikel, 1947–1958.
13. Personal records of Dr.-Ing. Heinz Nedelmann, Essen.
    Documentary fragments
14. Records located at the North Rhine Westphalian Main State Archive, Hauptstaatsarchiv Nordrhein-Westfalen (HStA NRW), Düsseldorf.
    NW72. Akten des Ministeriums für Landplanung, Wohnungen und öffentliche Arbeit
    NW75. Akten des Wirtschaftsministeriums, NRW. Kontrolle der deutschen Wirtschaft 1946–1955
    NW78. Akten des WiMins, NRW. Energiewirtschaft 1946–1956
    NW99. Akten des WiMins, NRW. Reparationen 1946–1961
    NW203. Akten des WiMins, NRW. Sondergruppe Demontage, bzw. Gruppe Reparationen

15. Records located at the Rhenish-Westphalian Economic Archive, Rheinisch-Westfälisches Wirtschaftsarchiv (RWWA), Cologne.
    Abt.1. Akten der Industrie- und Handelskammer zu Köln, 1945–1952
    Geschäftsberichte-Sammlung
16. City Archives of Ludwigshafen, Stadtarchiv, Ludwigshafen am Rhein.
    Various manuscripts
    Newspaper clipping collection
    "Wort und Zahl" (Hrsg. von Statistischem Amt der Stadt)
    Geschäftsberichte BASF. 1952–1954
    Rheinpfalz. 29.IX.45–31.XII.46
    Stadt-Anzeiger. 12.V.45–6.II.46
    "Amtliche Bekanntmachungen der Stadt Ludwigshafen"
    Die BASF
17. Records located at the Chemical Industry Trade Association, Archiv des Verbandes der Chemischen Industrie, Frankfurt.
    Trade association *Geschäftsberichte*. 1947–1952
    Wirtschaftsverband Chemische Industrie (Br. Zone). "Mitteilungsblatt." 1946 ff.

France

Archives of the French Foreign Ministry, Republique Française, Ministère des Affaires Etrangères, Archives et Documentation, Paris.
    C96. Archives Tripartites. Groupe de contrôle de l'I.G. Farben [located at Colmar, seen in Paris]
    C98. Archives de la Haut-Commissariat de la Republique Française en Allemagne. (a) Division Production Industrielle de la Direction General des Affaires Economiques et Financières; (b) Commisariat pour le Land Rhénanie-Palatinat [located at Colmar, seen in Paris]
    Série Y, 1944–1949. Archives Diplomatiques. Allemagne
    Série Z, 1944–1949. Allemagne
    Ministère des Affaires Etrangères. Affaires Allemands et Autrichiennes. "Imprimés, 1946–1954"
    Bulletin mensuel du Commandement en Chef Français en Allemagne. 1945–1949

Great Britain

Records located at the Public Record Office (PRO), Kew.
    AY7.   Fuel Research Station: Registered Files.

Selected Bibliography

BT11. Board of Trade. Commercial Dept. Correspondence and papers. 1866–1969
BT195. Board of Trade. Economic Affairs Office. 1947–1952
BT211. Board of Trade. German Division Files. 1944–1953
FO371. Foreign Office. General Correspondence
FO935. Control Office. Intelligence Objectives Subcommittee. 1933–1945
FO936. Control Office. Establishment
FO938. Control Office. Private Office Papers. 1945–1956
FO942. Control Office. Economic and Industrial Planning Staff. 1943–1947
FO943. Control Office. Economic 1943–1955
FO1005. Control Commission for Germany (CCG, BE) [differs from "Control Office"]. Records Library. 1943–1959
FO1013. CCG. NRW Region
FO1025. CCG. BICO. Secretariat
FO1028. CCG. Coal control
FO1031. CCG. HQ T-Force and FIAT
FO1034. CCG. Economic Subcommittee. 1945–1948
FO1036. CCG. Office of the Economic Adviser
FO1039. CCG. Economics Division
T236. Treasury. Overseas Finance Division

United States

1. Records located at the main building of the U.S. National Archives and Records Service (NA), Washington, DC.
    Record Group (RG) 43. Allied Control Council
    RG43. European Advisory Commission: Records of Philip E. Mosely, U.S. Political Adviser on Germany 1943–1945
    RG59. Office of Strategic Services, Research and Analysis Branch
    RG165. War Department, General and Special Staffs
    RG218. U.S. Joint Chiefs of Staff
    RG243. U.S. Strategic Bombing Survey (European War)
    RG353. Interdepartmental and Intradepartmental Committees (State Department)
2. Records located in the National Archives General Archives Division, Washington National Records Center (WNRC), Suitland, MD.
    RG84. U.S. Political Adviser's Office to Military Government in Germany
    RG169. Foreign Economic Administration, Enemy Branch
    RG260. U.S. Occupation Headquarters, World War II, Office of Military Government for Germany, U.S. (OMGUS)

RG319. Army Staff, Intelligence (G-2)
3. University of Illinois Archives, Urbana-Champaign, IL.
Roger Adams Papers

Interviews

Professor Dr. Kurt Hansen. Bayer A.G., Leverkusen. December 4, 1984
Dr.-Ing. Heinz Nedelmann. Essen. July 9, 1984
Dr. A. Felix Prentzel. Frankfurt. July 17, 1986

PRIMARY PRINTED WORKS

Adenauer, Konrad. *Memoirs 1945–1953.* Beate Ruhm v. Oppen, trans. Chicago: Henry Regnery, 1965.
Akten zur Vorgeschichte der Bundesrepublik Deutschland 1945–1949. 5 vols. Bundesarchiv & Institut für Zeitgeschichte, Hrsg. Munich/Vienna: R. Oldenbourg, 1976–1983.
Alphand, Hervé. *L' étonnement d'être. Journal 1939–1973.* Paris: Librairie Arthème Fayard, 1977.
Amelunxen, Rudolf. *Ehrenmänner und Hexenmeister. Erlebnisse und Betrachtungen.* Munich: Günter Olzog, 1960.
Angewandte Chemie. Vols. 59–61 (1947–1949).
Badische Anilin- und Soda-Fabrik, A.G. "Die Neugründung der BASF im Zuge der IG-Entflechtung und die Entwicklung der BASF." Erstattet vom Dr. Carl Wurster. Ludwigshafen, July 15, 1954.
Bailey, Thomas A. *The Marshall Plan Summer. An eye-witness report on Europe and the Russians in 1947.* Stanford, CA: Hoover Institution Press, 1977.
Basisdaten. Zahlen zur sozio-ökonomischen Entwicklung der Bundesrepublik Deutschland. Roland Ermrich, Bearb. Bonn-Bad Godesberg: Verlag Neue Gesellschaft, 1973.
Bayer A.G. *Revolution im Unsichtbaren.* Düsseldorf: Econ-Verlag, 1963.
Borkin, Joseph, and Charles A. Welsh. *Germany's Master Plan: The Story of Industrial Offensive.* New York: Duell, Sloan and Pearce, 1943.
Brown, Lewis H. *A Report on Germany.* New York: Farrar, Straus & Co., 1947.
Bundesminister für den Marshallplan. *Neunter Bericht der Deutschen Bundesregierung über die Durchführung des Marshallplans* (1. Okt. 1951 bis 31. Dez. 1951 und Jahresüberblick 1951). Bonn, n.p., 1952.
———, Hrsg. *OEEC Europäisches Wiederaufbauprogram. Zweiter Bericht.* Bonn, n.p., 1950.
Chandon, Emil C. *Die Industriestruktur des britischen Besatzungsgebietes (I).*

Essen/Ketwig: West-Verlag, 1947.
Chemie-Ingenieur-Technik. Vols. 19–23 (1947–1951). [Note that this is Teil B of *Angewandte Chemie* until 1950, when this journal changed its name.]
Chemiewirtschaft in Zahlen. Zusammengestellt vom Verband der Chemischen Industrie. 4. Aufl. Düsseldorf: Econ, 1960.
Clarke, Sir Richard. *Anglo-American Economic Collaboration in War and Peace 1942–1949.* Ed. by Sir Alec Cairncross. Oxford: Clarendon Press, 1982.
Clay, Lucius D. *Decision in Germany.* Garden City, NJ: Doubleday & Co., 1950.
———. *The Papers of General Lucius D. Clay.* 2 vols. Jean Edward Smith, ed. Bloomington, IN: Indiana Univ. Press, 1974.
Deutsches Institut für Wirtschaftsforschung. *Die deutsche Wirtschaft zwei Jahre nach dem Zusammenbruch.* Berlin: Albert Nauck, 1947.
"Deutsche Wirtschaft im Querschnitt. 10. Folge. Chemische Industrie." *Der Volkswirt,* Nr. 21 (25 May 1951): 17–52.
Dirlam, Joel and Alfred E. Kahn. *Fair Competition. The Law and Economics of Antitrust Policy.* Ithaca: Cornell University Press, 1954.
DuBois, Josiah E., Jr. *The Devil's Chemists: 24 Conspirators of the International Farben Cartel Who Manufacture Wars.* Boston: Beacon Press, 1952. [Published in Britain as *Generals in Grey Suits.* London: The Bodley Head, 1953.]
Duisberg, Curt. *Nur ein Sohn. Ein Leben mit der Grosschemie.* Stuttgart: Seewald, 1981.
Ehrmann, Felix. "Chemie und Industrie-Entwicklung in den letzten 25. Jahren." Frankfurt, n.p., 1954.
Europa-Archiv. 1946–1951.
European Recovery Program. *Second Annual Program (1949/50) for the United States and United Kingdom Occupied Areas in Germany.* N.p., n.d.
———. *Long-Term Program (1952–53) for the U.S. and U.K. Occupied Areas in Germany.* N.p., n.d.
Frederikson, Oliver J. *The American Military Occupation of Germany. 1945–1953.* Historical Division, United States Army, Europe, 1953.
Friedensburg, Ferdinand. *Die Rohstoffe und Energiequellen im neuen Europa.* Oldenburg/Berlin: Stallung, 1943.
Galbraith, John K. *American Capitalism. The Concept of Countervailing Power.* Boston: Houghton-Mifflin, 1952.
Gattineau, Heinrich. *Durch die Klippen des 20. Jahrhunderts. Erinnerungen zur Zeit- und Wirtschaftsgeschichte.* Stuttgart: Seewald, 1983.
Geiler, Kurt. *Geistige Freiheit und soziale Gerechtigkeit im neuen Deutschland.* Wiesbaden: Verlag der Greif, 1947.
Goudsmit, Samuel A. *Alsos.* Introduction by R. V. Jones. New York: Henry Schuman, 1947; Los Angeles: Tomash, 1983.

Gross, Hermann. *Ein Beitrag zur Aufteilung der I.G. Farbenindustrie Aktiengesellschaft.* Kiel: Institut für Weltwirtschaft (IfW), 1949.

———. *Bilanz der Forschungsausgaben in der amerikanischen Industrie.* Kiel: IfW, 1957.

———. *Material zur Aufteilung der I.G. Farbenindustrie Aktiengesellschaft.* Kiel: IfW, 1950.

———. *Untersuchung über die Kapitalstruktur repräsentativer Aktiengesellschaften der westdeutschen chemischen Industrie in den Jahren 1948-1950.* Kiel: IfW, 1952.

———. *Wirtschaftswichtige Forschung und Wissenschaftsfinanzierung im Deutschland und den USA.* Kiel: IfW, 1955.

Groves, Leslie R. *Now it can be told: The story of the Manhattan Project.* New York: Harper & Bros., 1962.

Handbuch der Grossunternehmen. 4. Aufl. Band: *Bundesrepublik Deutschland und Westberlin.* Darmstadt: Hoppenstedt & Co., 1955.

Härtel, Lia, Zusammensteller. *Der Länderrat des amerikanischen Besatzungsgebietes.* Stuttgart: Kohlhammer, 1951.

Heintzeler, Wolfgang. *Der rote Faden. Fünf Jahrzehnte. Staatsdienst. Wehrmacht, Chemische Industrie, Nürnberg, Marktwirtschaft, Mitbestimmung, Kirche.* Stuttgart: Seewald, 1983.

Hoechst A.G., Hrsg. *Dokumente aus Hoechst-Archiven.* Bde. 48–50. Frankfurt: Hoechst A.G., 1976–1978: Bd. 48: *US Administration. Die Verwaltung des Werkes Hoechst 1945–1953* (1976); Bd. 49: *Der Hoechst-Konzern entsteht. Die Verhandlungen über die Auflösung von IG-Farben und die Gründung der Farbwerke Hoechst A.G. 1945–1953*, Teil 1 (1978); Bd. 50: Ibid., Teil 2 (1978).

Hoover, Calvin. *Memoirs of Capitalism, Communism, and Nazism.* Durham: Duke University Press, 1965.

I.G. Farbenindustrie A.G. in Auflösung. *Bericht zur RM-Schlussbilanz zum 20. Juni 1940 und DM-Eröffnungsbilanz zum 21. Juni 1948.* N.d., Tripartite I.G. Farben Control Group.

I.G. Farbenindustrie A.G. *Werksgeschichte. Die Gefolgschaft der Werke Leverkusen, Elberfeld und Dormagen zur Erinnerung an die 75. Wiederkehr des Gründungstages der Farbenfabriken vorm. Friedr. Bayer & Co.* München: Graphischen Kunstanstalten F. Bruchmann, 1938.

———in Abwicklung. *Geschäftsberichte.* 1980–1983.

Die Kabinettprotokolle der Bundesregierung. Ulrich Enders & Konrad Reiser, Bearb. Bd. 1: 1949. Boppard: Harold Boldt, 1982.

Keesings Archiv der Gegenwart. Vols. 15–19 (1945–1949).

Knierheim, August v. *Nürnberg. Rechtliche und menschliche Probleme.* Stuttgart: Ernst Klett, 1953.

Krischan, Alexander, Bearb. *Bibliographie der I.G. Farben-Entflechtung.*

Darmstadt: Hoppenstedt, 1957.

Kropatt, Wolf-Arno. *Hessen in der Stunde Null 1945/1947. Politik. Wirtschaft und Bildungswesen in Dokumenten.* Wiesbaden: Selbstverlag der Historischen Kommission für Nassau, 1979.

Länderrat des Amerikanischen Besatzungsgebietes. *Statistisches Handbuch von Deutschland 1928–1944.* München: Franz Ehrenwirth, 1949.

Lanz, Kurt. *Weltreisender in Cheme.* Düsseldorf: Econ, 1978. [English: *Around the World with Chemistry.* New York: McGraw-Hill, 1980.]

Leitende Männer der Wirtschaft und der zugehörigen Verwaltung. Ein wirtschaftliches "Who is who?" Nachschlagswerk über Vorstandsmitglieder, Aufsichtsräte u.s.w. 1951. Heppenheim (Bergstr.): Hoppenstedt, 1951.

Lilienthal, David E. *Big Business: A New Era.* New York: Harper, 1953.

Martin, James. *All Honorable Men.* Boston: Little, Brown, 1950.

Massigli, René. *Une comédie des erreurs, 1943–1956: Souvenirs et réflexions sur une étape de la construction européene.* Paris: Plon, 1978.

ter Meer, Fritz. *Die I.G. Farbenindustrie Aktiengesellschaft. Ihre Entstehung, Entwicklung und Bedeutung.* Düsseldorf: Econ, 1953.

Morgenthau, Henry, Jr. *Morgenthau Diary (Germany).* 2 vols. Washington: USGPO, 1967.

Murphy, Robert. *Diplomat among Warriors.* Garden City, NJ: Doubleday & Co., 1964; Pyramid Books ed., 1965.

Oppen, Beate Ruhm von, ed. *Documents on Germany under Occupation. 1945–1954.* London: Oxford University Press, 1955.

Pünder, Hermann. *Von Preussen nach Europa. Lebenserinnerungen.* Stuttgart: Deutsche Verlags-Anstalt, 1968.

Ratchford, B. U., and William Ross. *Berlin Reparations Assignment.* Chapel Hill: University of North Carolina Press, 1947.

Sasuly, Richard. *I.G. Farben.* New York: Boni & Gaer, 1947.

Seydoux, Francois. *Memoires d'outre-Rhin.* Paris: Bernard Grasset, 1975.

Simon, Leslie E. *German Research in World War II.* New York: John Wiley, 1947.

Speer, Albert. *Inside the Third Reich.* New York: MacMillan, 1970; Avon Books ed. 1976.

Statistisches Bundesamt, Hrsg. *Bevölkerung und Wirtschaft, 1872–1972.* Stuttgart: Kohlhammer, 1972.

———. *Die Industrie der BRD. Reihe 4: Sonderveröffentlichungen. Heft 17: Die Industrie in den europäischen Ländern bis 1956. Betriebe, Beschäftigung und Produktionswerte nach Industriezweigen. Länderübersichten.* Stuttgart: Kohlhammer, 1958.

Statisches Landesamt Nordrhein-Westfalen, Hrsg. *Chemische-Industrie und Mineralölverarbeitung in Nordrhein-Westfalen 1960 u. 1965.* Düsseldorf: Statistisches Landesamt NRW, 1966.

Supreme Headquarters, Allied Expeditionary Force, Office of the Chief of Staff. *Handbook for Military Government in Germany prior to defeat or surrender.* N.p., December 1944.

Trials of War Criminals before the Nuernberg Military Tribunals under Control Council Law No. 10. Vols. VII and VIII: "The Farben Case." Washington: USGPO, 1952-1953.

U.S. Department of State. "Accord on Treatment of German-owned Patents." July 27, 1946. *Treaties and other international acts series*, no. 2415. Washington: USGPO, n.d.

―――, *Foreign Relations of the United States [FRUS]*, 1945, Vol. III: *The European Advisory Commission; Austria; Germany.* Washington: USGPO, 1968.

―――. *FRUS*, 1946, Vol. II: *Council of Foreign Ministers.* Washington: USGPO, 1970.

―――. *FRUS*, 1946, Vol. V: *The British Commonwealth: Western and Central Europe.* Washington: USGPO, 1969.

―――. *FRUS*, 1947, Vol. II: *Council of Foreign Ministers: Germany and Austria.* Washington: USGPO, 1972.

―――. *FRUS*, 1948, Vol. III: *Western Europe.* Washington: USGPO, 1974.

―――. *FRUS*, 1949, Vol. III: *Council of Foreign Ministers: Germany and Austria.* Washington: USGPO, 1974.

―――. *United States Economic Policy toward Germany.* Washington: USGPO, 1946.

―――. *Germany 1947-1949: The Story in Documents.* Washington: USGPO, 1950.

U.S. Economic Cooperation Administration. *Western Germany Country Study.* Washington: USGPO, 1949.

U.S. Foreign Economic Administration. *Report to Congress on Operations.* Washington: USGPO, 1944.

U.S. High Commission for Germany. Econ. Rpts. Br. *Annual Industries Rpt. 1952: Federal Republic of Germany and Western Sectors of Berlin, Part V: Chemicals and Rubber.* N.p., September 16, 1953.

U.S. Office of Military Government for Germany. *Control of I.G. Farben.* Special Report to the Governor of the U.S. Zone. October 1, 1945.

―――. (OMGUS) *Ermittlungen gegen die I.G. Farbenindustrie A.G.— September 1945.* (Trans. and comp. by Dokumentenstelle zur NS-Sozialpolitik, Hamburg.) Nördlingen: Franz Greno, 1985.

―――. *Monthly Report of the Military Governor, U.S. Zone.* 1945-1949.

―――, Economics Division. *Economic Data on Potsdam Germany.* Special Report of the Military Governor. September 1947.

―――. *A Year of Potsdam.* 1946.

Selected Bibliography

U.S. Strategic Bombing Survey (European War). USSBS Rpt. 115. *Ammoniakwerke Merseburg GmbH, Leuna, Germany.* 2d ed. March 1947.

———. USSBS Rpt. 3: *The Effects of Strategic Bombing on the German War Economy.* October 31, 1945.

———. USSBS Rpt. 117. *Ludwigshafen-Oppau Works of I.G. Farbenindustrie A.G., Ludwigshafen, Germany.* 2d ed. January 1947.

———. USSBS Rpt. 109. *Oil Division Final Report.* 2d ed. January 1947.

———. USSBS Rpt. 110. *Oil Division Final Report Appendix.* 2d ed. January 1947.

———. USSBS Rpt. 2. *Overall Report.* September 30, 1945.

Vaubel, Ludwig. *Zusammenbruch und Wiederaufbau. Ein Tagebuch aus der Wirtschaft.* München: R. Oldenbourg, 1984.

Wer leitet? Die Männer der Wirtschaft und der einschlägigen Verwaltung 1940. Berlin: Hoppenstedt & Co., 1940.

———. *1941/42.* Berlin: Hoppenstedt & Co., 1942.

White, Theodore H. *Fire in the Ashes: Europe in Mid-Century.* New York: William Sloane Associates, 1953.

Wickel, Helmuth. *I.G. Deutschland. Ein Staat im Staate.* Berlin: Verlag Der Bücherkreis, 1932.

Die Wiedergesundung Europas. Schlussbericht der Pariser Wirtschaftskonferenz der sechzehn Nationen. Two parts. Oberursel (Taunus): Verlag Europa-Archiv, 1948.

Wiel, Paul. "Die Industrie der Nichtkohlechemie und der NE-Metalle im Ruhrgebiet." Pp. 7–30 of Paul Wiel and others. *Beiträge zur Industriewirtschaft des Ruhrgebietes.* Essen: West-Verlag, 1947.

Winkelmeyer, Gregor. "Standortsfragen der Kohlechemie des Ruhrgebietes." Pp. 31–51 of ibid.

Winnacker, Karl. *Nie den Mut verlieren. Erinnerungen an Schicksaljahren der deutschen Chemie.* Düsseldorf: Econ, 1971.

Wirtschaftsstatistik der deutschen Besatzungszonen 1945–1948 in Verbindung mit der deutschen Produktionsstatistik der Vorkriegszeit. Oberursel: Europa-Archiv, 1948.

Wörtliche Berichte und Drucksachen des Wirtschaftsrats des Vereinigten Wirtschaftsgebiets 1947–1949. Hrsg. vom Institut für Zeitgeschichte und dem Deutschen Bundestag Wiss.-Dienste. Bearb. Christoph Weisz und Hans Woller. 6 vols. Munich: Oldenbourg, 1977.

SECONDARY SOURCES

Abelshauser, Werner. "The First Post-Liberal Nation: Stages in the Development of Modern Corporatism in Germany." *European History Quarterly (EHQ)* 14 (July 1984): 285–318.

———. "Freiheitlicher Sozialismus oder soziale Marktwirtschaft? Die Gutachtertagung über Grundfragen der Wirtschaftsplanung am 21. und 22. Juni 1946." *Vierteljahrshefte für Zeitgeschichte (VfZ)* 24 (October 1976): 415–449.

———. "Korea, die Ruhr und Erhards Marktwirtschaft. Die Energiekrise von 1950–1951." *Rheinische Vierteljahrsblätter* 45 (1981): 287–316.

———. "Von der Kohlenkrise zur Gründung der Ruhrkohle A.G." Pp. 415–443 of Hans Mommsen and Ulrich Borsdorf, eds. *Glück auf, Kameraden! Die Bergarbeiter und ihre Organisationen in Deutschland*. Köln; Bund, 1978.

———. *Der Ruhrkohlenbergbau seit 1945. Wiederaufbau, Krise, Anpassung*. München: C. H. Beck, 1984.

———. "Wiederaufbau vor dem Marshall-Plan. Westeuropas Wachstumschancen und die Wirtschaftsordnungspolitik in der zweiten Hälfte der vierziger Jahre." *VfZ* 29 (October 1981): 545–578.

———. *Wirtschaft in Westdeutschland 1945–1948. Rekonstruktion und Wachstumsbedingungen in der amerikanischen und britischen Zone*. Stuttgart: Deutsche Verlags-Anstalt, 1975.

———. "Wirtschaft und Besatzungspolitik in der französischen Zone, 1945–1949." Pp. 111–139 of Claus Scharf and Hans-Jürgen Schröder, eds. *Die Deutschlandpolitik Frankreichs und die französischen Zone 1945–1949*. Wiesbaden: Franz Steiner, 1983.

———. *Wirtschaftsgeschichte der Bundesrepublik Deutschland 1945–1980*. Frankfurt: Suhrkamp, 1983.

Adamsen, Heiner R. "Faktoren und Daten der wirtschaftlichen Entwicklung in der Frühphase der Bundesrepublik 1948–1954." *Archiv für Sozialgeschichte (AfS)* 18 (1978): 217–244.

Amrosius, Gerold. "Funktionswandel und Strukturveränderung der Bürokratie 1945–1949. Das Beispiel der Wirtschaftsverwaltung." Pp. 167–207 of Heinrich A. Winkler, ed. *Politische Weichenstellung im Nachkriegsdeutschland 1945–1953*. Göttingen: Vandenhoeck & Ruprecht, 1979.

Backer, John H. *Winds of History. The German Years of Lucius DuBignon Clay*. New York: Van Nostrand Reinhold, 1983.

———. *Priming the German Economy*. Durham, NC: Duke University Press, 1971.

Balabkins, Nicholas. *Germany under Direct Controls: Economic Aspects of Industrial Disarmament 1945–1949*. Brunswick, NJ: Rutgers University Press, 1964.

Barker, R. S. "Civil Service Attitudes and Economic Planning of the Attlee Government." *Journal of Contemporary History* 21 (July 1986): 473–486.

Barnikel, Hans-Heinrich. "Die Konzentrationspolitik nach 1945." Pp.

54–73 of Hans Pohl and Wilhelm Treue, eds. *Die Konzentration in der deutschen Wirtschaft seit dem 19. Jahrhundert.* Wiesbaden: Steiner, 1978.

Bäumler, Ernst. *Ein Jahrhundert Chemie.* Düsseldorf: Econ, 1963.

Bayer A.G. *Beiträge zur hundertjährigen Firmengeschichte 1863–1963.* Leverkusen: Bayer A.G., 1963/1964.

Beer, John J. *The Emergence of the German Dye Industry.* Urbana: University of Illinois Press, 1959.

Benz, Wolfgang. "Zwangswirtschaft und Industrie. Das Problem der Kompensationsgeschäfte am Beispiel der Kasseler Spinnfaser-Prozesses von 1947." *VfZ* 32 (July 1984): 442–440.

Berghahn, Volker. "Deutschland, Amerika und die 'Neuordnung' der Weltwirtschaft 1933–1960." *Neue politische Literatur (NpL)* 29 (1984): 335–350.

———. *The Americanisation of West German Industry, 1945–1973.* Cambridge/New York: Cambridge University Press, 1986.

———. "Westdeutsche Unternehmer, Weltmarkt und Wirtschaftsordnung: Zur Bedeutung des Kartellgesetzes." Pp. 301–324 of Lothar Albertin and Werner Link, eds. *Politische Parteien auf dem Weg zur parlamentarischen Demokratie in Deutschland.* Düsseldorf: Droste, 1981.

Beyerchen, Alan. "German Scientists and Research Institutions in Allied Occupation Policy." *History of Education Quarterly* 22 (Fall 1982): 289–299.

Bibliographie zur Deutschlandpolitik 1975–1982. Hrsg. vom Bundesministerium für Innerdeutsche Beziehungen. Bearb. von Karsten Schröder. Frankfurt: Alfred Metzger, 1983.

Birkenfeld, Wolfgang. *Der synthetische Treibstoff 1933–1945. Ein Beitrag zur nationalsozialistischen Wirtschafts- und Rüstungspolitik.* Göttingen: Musterschmidt, 1964.

Borden, William S. *The Pacific Alliance: United States Foreign Economic Policy and Japanese Trade Recovery, 1947–1955.* Madison: University of Wisconsin Press, 1984.

Borkin, Joseph. *The Crime and Punishment of I.G Farben.* New York: Free Press, 1978; Pocket Books ed., 1979.

Botting, Douglas. *In the Ruins of the Reich.* London: George Allen & Unwin, 1985.

Braunthal, Gerald. "The Anglo-Saxon Model of Democracy in the West German Political Consciousness after World War II." *AfS* 18 (1978): 245–277.

Bullock, Alan. *Ernest Bevin: Foreign Secretary, 1945–1951.* London: Heinemann, 1983.

Carden, Robert W. "Before Bizonia: Britain's Economic Dilemma in Germany, 1945–1946." *Journal of Contemporary History* 14 (July 1979): 535–555.

Carr, William. *Arms, Autarky and Aggression.* New York: W. W. Norton, 1972.
Castillon, Richard. *Les réparations allemandes. Deux expériences 1919–32, 1945–52.* Paris: Presse Univ. de France, 1953.
Chabir, Gabriele, and Michael Haupt. "Die Teilung Deutschlands 1945–1949. Bericht und Bibliographie." Pp. 359–390 of *Jahresbibliographie 1977. Bibliothek für Zeitgeschichte.* München: Bernard und Graefe Verlag für Wehrwesen, 1978.
Czichon, Eberhard. *Der Bankier und die Macht. Hermann Josef Abs in der deutschen Politik.* Köln: Pahl-Ruggenstein, 1970.
———. *Hermann Josef Abs. Porträt eines Kreuzritters des Kapitals.* Berlin (O): Union Verlag, 1969.
Danek, Paul. "Zur reaktionären Rolle des staatsmonopolistischen Kapitalismus bei der Wiedererrichtung und Machtausweitung des I.G.-Farbenmonopols in Westdeutschland." Dissertation, Halle, 1961.
Daniel, Ute. *Dollardiplomate in Europa. Marshallplan, Kalter Krieg und U.S.-Aussenwirtschaftspolitik.* Düsseldorf: Droste, 1982.
Dapper, Karl-Peter and Gerhard Hahn, eds. *Bibliographien zur Sozialen Marktwirtschaft. Die Wirtschafts- und Gesellschaftsordnung der Bundesrepublik Deutschland 1945/49–1981.* Baden-Baden: Nomos Verlagsgesellschaft, 1983.
Denzel, Rosemarie. "Die chemische Industrie Frankreichs unter der deutschen Besatzung im zweiten Weltkrieg." Tübingen: Institut für Besatzungsfragen, 1959. [mimeo-typescript]
Deubner, Christian, Udo Rehlfedt and Frieder Schlupp. "Deutschfranzösische Wirtschaftsbeziehungen im Rahmen der weltwirtschaftlichen Arbeitsteilung. Interdependenz, Divergenz oder strukturelle Dominanz?" Pp. 91–136 of Robert Picht, ed. *Deutschland-Frankreich-Europa. Bilanz einer schwierigen Partnerschaft.* München: Piper, 1978.
Domes, Jürgen, and Michael Wolfsohn. "Setting the Course for the Federal Republic of Germany: Major Policy Decisions in the Bizonal Economic Council and Party Images, 1947–1949." *Zeitschrift für die gesamte Staatswissenschaft* (ZgS) 135 (September 1979): 332–351.
Dorn, Walter. "The Debate over American Occupation Policy in Germany in 1944–1945." *Political Science Quarterly* 72 (December 1957): 481–501.
Drexler, Alexander, Wolfgang Krumbein, and Friedrich Stratmann. "Die britischen 'Sparta-Pläne' 1946." Pp. 245–263 in Josef Foschepoth and Rolf Steininger, eds. *Die britische Deutschland- und Besatzungspolitik 1945–1949.* Paderborn: Schöningh, 1985.
Eckes, Alfred E., Jr. *A Search for Solvency: Bretton Woods and the International Monetary System 1941–1971.* Austin: University of Texas Press, 1975.
Eichholtz, Dietrich. "Die I.G.-Farben- 'Friedensplanung.' Schlüsseldo-

kumente der faschistischen 'Neuordung des europäischen Grossraums.'" *Jahrbuch für Wirtschaftsgeschichte (JfW)* (Teil III, 1966): 271–332.

———. "Zum Anteil des I.G.-Farbenkonzerns an der Vorbereitung des Zweiten Weltkrieges." *JfW* (Teil II, 1969): 89–105.

Eisenberg, Carolyn. "Working-Class Politics and the Cold War: American Intervention in the German Labor Movement, 1945–1949." *Diplomatic History* 7 (Fall 1983): 283–306.

Etzold, Heike. "Carl Duisberg—vom stellungsuchenden Chemiker an die Spitze der I.G. Farbenindustrie A.G." *JfW* (Teil III, 1966): 196–215.

Eulenburg, Franz. "Die Herkunft der deutschen Wirtschaftsführer." *Schmollers Jahrbuch für Gesetzgebung, Verwaltung und Volkswirtschaft* 74 (1954): 77–89.

Fachverband Kohlechemie und verwandte Gebiete, e.V. *Wichtige Zahlen. Teil I: Inland.* 23 Jg. 1971.

Fischer, Wolfram. "Dezentralisation oder Zentralisation—kollegiale oder autoritäre Führung? Die Auseinandersetzung um die Leitungsstruktur bei der Entstehung des I.G. Farbenkonzerns." Pp. 476–488 in Norbert Horn and Jürgen Kocka, eds. *Recht und Entwicklung der Grossunternehmen im 19. und frühen 20. Jahrhundert.* Göttingen: Vandenhoeck & Ruprecht, 1979.

Flechtner, Hans-Joachim. *Carl Duisberg. Vom Chemiker zum Wirtschaftsführer.* Düsseldorf: Econ, 1959.

Flora, Peter, Bearb. *Indikatoren der Modernisierung. Ein historisches Datenhandbuch.* Opladen: Westdeutscher Verlag, 1975.

Foelz-Schroeter, Marie Elise. *Föderalistische Politik und nationale Repräsentation 1945–1947.* Stuttgart: Deutsche Verlags-Anstalt, 1974.

Forster, Karl-Heinz. *Finanzierung durch Abschreibungen. Nach den Ergebnissen von DM-Bilanzen.* Stuttgart: C. E. Poeschel, 1953.

Foschepoth, Josef. "Britische Deutschlandpolitik zwischen Jalta und Potsdam." *VfZ* 30 (October 1982): 675–714.

———, ed. *Kalter Krieg und Deutsche Frage.* Göttingen: Vandenhoeck & Ruprecht, 1985.

———, and Rolf Steiniger, eds. *Die britische Deutschland- und Besatzungspolitik 1945–1949.* Paderborn: Schöningh, 1985.

Fröchte, Heribert. "Die wirtschaftliche Bedeutung der Konzernentflechtung." Dissertation, Frankfurt a.M., 1949.

Fromme, Friedrich. "Zur inneren Ordnung in den westlichen Besatzungszonen 1945–1949." *VfZ* 10 (April 1962): 206–223.

Gaddis, John Lewis. *The United States and the Origins of the Cold War. 1941–1947.* New York: Columbia University Press, 1972.

Gardner, Lloyd C. "America and the German 'Problem,' 1945–1949."

Pp. 113–148 of Barton J. Bernstein, ed. *Politics and Policies of the Truman Administration.* Chicago: Quadrangle, 1970.

Gardner, Richard N. *Sterling-Dollar Diplomacy. Anglo-American Collaboration in the Reconstruction of Multilateral Trade.* Oxford: Clarendon Press, 1956.

Gerbet, Pierre. "La Genèse du Plan Schuman." *Revue française de science politique* 6 (July–September 1956): 525–553.

Geschichte des VEB-Leuna-Werke "Walter Ulbricht" 1945 bis 1981. Leipzig: VEB Deutscher Verlag für Grundstoffindustrie, 1986.

Gillingham, John. "A Case of Continuity: The Cartelization of the Western European Montanindustrie and the European Coal and Steel Community." St. Louis, MO: University of Missouri at St. Louis, 1979.

———. *Industry and Politics in the Third Reich. Ruhr Coal, Hitler and Europe.* London: Methuen, 1985.

Gimbel, John. *The American Occupation of Germany.* Stanford: Stanford Univesity Press, 1968.

———. *The Origins of the Marshall Plan.* Stanford: Stanford University Press, 1976.

Gormley, James L. "The Washington Declaration and the 'Poor Relation': Anglo-American Atomic Diplomacy, 1945–1946." *Diplomatic History* 8 (Spring 1984): 125–143.

Greenwood Sean. "Ernest Bevin, France and 'Western Union': August 1945–February 1946." *EHQ* 14 (July 1984): 319–338.

Gressley, Gene M. "Thurman Arnold and the New Deal." *BHR* 38 (Summer 1964): 214–231.

Grosser, Alfred. *L'Allemagne de l'Occident, 1945–1952.* Paris: Gallimard, 1953.

Guradze, Heinz. "The *Länderrat*: Landmark of German Reconstruction." *Western Political Quarterly* 3 (1956): 190–213.

Haber, L. F. *The Chemical Industry during the Nineteenth Century: A Study of the Economic Aspects of Applied Chemistry in Europe and North America.* Oxford: The Clarendon Press, 1971.

———. *The Chemical Industry 1900–1930. International Growth and Technological Change.* Oxford: Clarendon Press, 1971.

———. *The Poisonous Cloud: Chemical Warfare in the First World War.* Oxford: Clarendon Press, 1986.

Hardach, Gerd. *Deutschland in der Weltwirtschaft 1870–1970. Eine Einführung in die Sozial- und Wirtschaftsgeschichte.* Frankfurt: Campus, 1977.

Hartmann, Peter C. "Die politische und wirtschaftliche Entwicklung Frankreichs im Zweiten Weltkrieg. Grundlage und Voraussetzung für die frühe französische Besatzungspolitk in Südwestdeutschland." Pp. 179–192 of Hans Schwarzmaier, ed. *Landesgeschichte und Zeitgeschichte. Kriegsende 1945 und demokratischer Neubeginn am Oberrhein.* Karlsruhe:

Kommissionsverlag G. Braun, 1980.
Hawley, Ellis W. *The New Deal and the Problem of Monopoly: A Study in Economic Ambivalence*. Princeton: Princeton University Press, 1966.
Hayes, Peter F. "Carl Bosch and Carl Krauch: Chemistry and the Political Economy of Germany, 1925-1945." *Journal of Economic History* 47 (June 1987): 353-363.
―――. *Industry and Ideology: I.G. Farben in the Nazi Era*. Cambridge/New York: Cambridge University Press, 1987.
Henke, Klaus-Dietmar. "Aspekte französischer Besatzungspolitik in Deutschland nach dem Zweiten Weltkrieg." Pp. 169-191 of *Miscellanea. Festschrift für Helmut Krausnick zum 75. Geburtstag*. Stuttgart: Deutsche Verlags-Anstalt, 1980.
―――. "Politik der Widersprüche. Zur Charakteristik der französischen Militärregierung in Deutschland nach dem Zweiten Weltkrieg." Pp. 49-85 of Claus Scharf and Hans-Jürgen Schröder, eds. *Die Deutschlandpolitik Frankreichs und die französische Zone 1945-1949*. Wiesbaden: Franz Steiner, 1983.
―――. *Politische Säuberung unter französischer Besatzung. Die Entnazifizierung in Württemberg-Hohenzollern*. Stuttgart: Deutsche Verlags-Anstalt, 1981.
Herbst, Ludolf. "Krisenüberwindung und Wirtschaftsneuordnung, Ludwig Erhards Beteiligung an den Nachkriegsplanungen am Ende des Zweiten Weltkrieges." *VfZ* 25 (1977): 305-340.
―――. *Der totale Krieg und die Ordung der Wirtschaft. Die Kriegswirtschaft im Spannungsfeld von Politik, Ideologie und Propaganda 1939-1945*. Stuttgart: Deutsche Verlags-Anstalt, 1982.
Herzstein, Robert E. *When Nazi Dreams Come True: The Third Reich's Internal Struggle over the Future of Europe after a German Victory. A Look at the Nazi Mentality, 1939-1945*. London: Abacus, 1982.
Hilberg, Raul. *The Destruction of the European Jews*. Chicago: Quadrangle, 1961; Quadrangle Paperbacks ed., 1967.
Hillel, Marc. *L'occupation française en allemagne (1945-1949)*. Paris: Balland, 1983.
Hoffmann, W. G., and others. *Das Wachstum der deutschen Wirtschaft seit der Mitte des 19en Jahrhunderts*. Berlin: Springer, 1965.
Hofmann, Rolf. *Weltchemiewirtschaft. Entwicklungstendenzen*. Opladen: Westdeutscher Verlag, 1975.
Hogan, Michael J. "American Marshall Planners and the Search for a European Neocapitalism." *AHR* 90 (February 1985): 44-72.
―――. "Paths of Plenty: Marshall Planners and the Debate over European Integration, 1947-1949." *Pacific Historical Review* 53 (February 1984). 337-366.
―――. "Revival and Reform: America's Twentieth-Century Search for a

New Economic Order Abroad." *Diplomatic History* 8 (Fall 1984): 287–310.

———. "The Search for a 'Creative Peace': The United States, European Unity, and the Origins of the Marshall Plan." *Diplomatic History* 6 (Summer 1982): 267–285.

Hohenberg, Paul M. *Chemicals in Western Europe, 1850–1914: An Economic Study of Technical Change.* Chicago: Rand McNally, 1967.

Hordemann, K. O. "L'évolution des frais de transport pour le charbon allemand et le charbon américain et leurs effets sur la position concurrientielle dans l'allemagne fédérale." Pp. 3–39 of *Les transports d'énergie. Travaux du Colloque Européen d'Economie de l'Energie. Grenoble, 6–8.V.1965.* Paris: La Haye, 1965.

Horn, Manfred. *Die Energiepolitik der Bundesregierung von 1958 bis 1972.* Berlin: Duncker & Humblot, 1977.

Hudemann, Rainer. "Französische Besatzungspolitik 1945–1952." *Neue politische Literatur* 26 (1981): 325–360.

———. "La zone francaise d'occupation sous le premier gouvernement du Général de Gaulle (mai 1945 a janvier 1946)." *Etudes gaulliennes* 6 (1978): 25–37.

Hughes, Thomas Parke, "Technological Momentum in History: Hydrogenation in Germany 1898–1933." *Past and Present* 44 (August 1969): 106–132.

Huppert, Walter. *Industrieverbände. Organisation und Aufgaben. Probleme und neue Entwicklungen.* Berlin: Duncker & Humblot, 1973.

Huster, Ernst-Ulrich, and others (Autorenkollektiv). *Determinanten der westdeutschen Restauration 1945–1949.* Frankfurt: Suhrkamp, 1972.

I.G. Chemie-Papier-Keramik, Hrsg. "Wilhelm Gefeller—sein Leben und sein Werk." By Brigit Hormann-Reckeweg and Norbert Weinitschke. Hanover, 1983.

Jerchow, Friedrich. *Deutschland in der Weltwirtschaft 1944–1947. Alliierte Deutschland- und Reparationspolitik und die Anfänge der westdeutschen Aussenwirtschaft.* Düsseldorf: Droste, 1978.

Kaufman, Burton. "Oil and Antitrust: The Oil Cartel Case and the Cold War." *Business History Review* 51 (1977): 35–61.

Kimball, Warren F. *Swords or Plowshares? The Morgenthau Plan for defeated Nazi Germany 1943–1946.* Philadelphia: J. B. Lippincott, 1976.

Klein, Burton H. *Germany's Economic Preparations for War.* Cambridge, MA: Harvard University Press, 1959.

Knapp, Manfred, ed. *Die deutsch-amerikanischen Beziehungen nach 1945.* Frankfurt: Campus, 1975.

———. "Reconstruction and West Integration: The Impact of the Marshall Plan on Germany." *ZgS* 137 (September 1981): 415–433.

Kocka, Jürgen. "1945: Neubeginn oder Restauration?" Pp. 141–168 in Carola Stern and Heinrich A. Winkler, eds. *Wendepunkte deutscher Geschichte 1848–1945.* Frankfurt: Fischer, 1979.

———, and Hannes Siegrist. "Die hundert grössten deutschen Unternehmen im späten 19. und frühen 20. Jahrhundert. Expansion, Diversifikation und Integration im internationalen Vergleich." Pp. 55–122 of Norbert Horn and Jürgen Kocka, eds. *Recht und Entwicklung der Grossunternehmen im 19. und 20. Jahrhundert.* Göttingen: Vandenhoeck & Ruprecht, 1979.

Korff, Adalbert. *Le revirement de la politique française a l'égard de l'Allemagne entre 1945 et 1950.* Ambilly-Annemasse: Imprimerie France-Suisse, 1965.

Kovaleff, Theodore. *Business and Government during the Eisenhower Administration: A Study of the Antitrust Division of the Justice Department.* Athens, OH: Ohio University Press, 1980.

Krammer, Arnold. "Fueling the Third Reich." *Technology and Culture* 19 (July 1978): 394–422.

———. "Technology Transfer as War Booty: The U.S. Technical Oil Mission in Europe, 1945." *Technology and Culture* 22 (January 1981): 68–103.

Kreikamp, Hans-Dieter, "Die Entflechtung der I.G. Farbenindustrie A.G. und die Gründung der Nachfolgegesellschaften." *VfZ* 25 (1977): 220–251.

Krengel, Rolf. *Anlagevermögen, Produktion und Beschäftigung der Industrie im Gebiet der Bundesrepublik von 1924 bis 1956.* Berlin: Duncker & Humblot, 1958.

Krieger, Wolfgang. "Was General Clay a Revisionist? Strategic Aspects of the United States Occupation of Germany." *Journal of Contemporary History* 18 (1983): 165–184.

Kuklick, Bruce. *American Policy and the Division of Germany: The Clash with Russia over Reparations.* Ithaca: Cornell University Press, 1972.

———. "The Genesis of the European Advisory Commission." *Journal of Contemporary History* 4 (October 1969): 189–201.

Kurz, Hans Otto. "Die technische, wirtschaftliche, politisch-rechtliche Entwicklung der chemischen Industrie Deutschlands seit 1900." Dissertation, Heidelberg, 1948.

Lange-Quassowski, Jutta-B. *Neuordnung oder Restauration.* Opladen: Leske Verlag & Budrick GmbH, 1979.

Latour, Conrad F., and Thilo Vogelsang. *Okkupation und Wiederaufbau. Die Tätigkeit der Militärregierung in der amerikanischen Besatzungszone Deutschlands 1944–1947.* Stuttgart: Deutsche Verlags-Anstalt, 1973.

Lattard, Alain. "Gewerkschaften und Betriebsräte in Rheinland-Pfalz 1945–1947. Zur französischen Gewerkschaftspolitik in Deutschland nach dem Zweiten Weltkrieg." Pp. 155–184 of Claus Scharf and Hans-Jürgen Schröder, eds. *Die Deutschlandpolitik Frankreichs und die französischen Zone 1945–1949*, Wiesbaden: Steiner, 1983.

Lauerson, Walter. "Mineralöl als Chemierohstoff. Ein Beitrag zur Würdigung einer neuen Entwicklung der chemischen Industrie." Kiel: Institut für Weltwirtschaft, 1952.

Laufer, Rudolf. *Industrie und Energiewirtschaft im Land Baden 1945–1952. Südbaden unter französischer Besatzung*. Freiburg: Karl Alber, 1979.

———. "Die südbadische Industrie unter französischer Besatzung 1945–1949." Pp. 141–153 of Claus Scharf and Hans-Jürgen Schröder, eds. *Die Deutschlandpolitik Frankreichs und die französische Zone 1945–1949*. Wiesbaden: Steiner, 1983.

Lefebure, Victor. *The Riddle of the Rhine: Chemical Strategy in Peace and War*. New York: The Chemical Foundation, 1923.

"Die Legende von der 'Stunde Null.' Planungen 1940–1950." Sonderheft of *Stadtbauwelt* 84 Ausgabe A (December 28, 1984).

Lerner, Franz. *Wirtschafts- und Sozialgeschichte des Nassauer Raumes 1816–1964*. Wiesbaden: N.p., 1964.

Lesch, Manfred. *Die Rolle der Offiziere in der deutschen Wirtschaft nach dem Ende des Zweiten Weltkrieges*. Berlin: Duncker & Humblot, 1970.

Link, Werner. *Deutsche und amerikanische Gewerkschaften und Geschäftsleute 1945–1975. Eine Studie über transnationale Beziehungen*. Düsseldorf: Droste, 1978.

———. "Der Marshall-Plan und Deutschland." *Aus Politik und Zeitgeschichte* B50 (1980): 3–18.

———. "Zum Problem der Kontinuität der amerikanischen Deutschlandpolitik im zwanzigsten Jahrhundert." Pp. 86–131 of Manfred Knapp, ed. *Die deutsch-amerikanischen Beziehungen nach 1945*. Frankfurt: Campus, 1975.

Lochner, Louis P. *Tycoons and Tyrant. German Industry from Hitler to Adenauer*. Chicago: Henry Regnery, 1954.

Loth, Wilfried. "Frankreich und die europäische Einigung." *Francia* 3 (1975): 699–705.

———. "Die Franzosen und die deutsche Frage 1945–1949." Pp. 27–48 of Claus Scharf and Hans-Jürgen Schröder, eds. *Die Deutschlandpolitik Frankreichs und die französischen Zone 1945–1949*. Wiesbaden: Steiner, 1983.

Lüders, Carsten. "Die Regelung der Ruhrfrage in den Verhandlungen über die politische und ökonomische Stabilisierung Westdeutschlands 1947–1949." Pp. 87–103 of Dietmar Petzina and Walter Euchner, eds. *Wirtschaftspolitik im britischen Besatzungsgebiet 1945–1949*. Düsseldorf: Schwann, 1984.

Lynch, Frances. "French Reconstruction in a European Context." European University Institute (Florence), EUI Working Papers No. 86 (1984).

———. "Resolving the Paradox of the Monnet Plan: National and International Planning in French Reconstruction." *Economic History Review* 2d Series, 37 (May 1984): 229–243.

McCraw, Thomas K. *Prophets of Regulation.* Cambridge, MA: Belknap Press of Harvard University Press, 1984.

Mai, Gunther. "Kontinuität und Neubeginn in der 'Stunde Null.' Einige Neuerscheinungen zur Regional- und Ortsgeschichte Westdeutschlands 1945/46." *Hessisches Jahrbuch für Landesgeschichte* 31 (1981): 231–256.

Maier, Charles S. "The Politics of Productivity: Foundations of American International Economic Policy after World War II." Pp. 23–49 of Peter Katzenstein, ed., *Between Power and Plenty: The Foreign Economic Policies of Advanced Industrial States.* Madison: University of Wisconsin Press, 1978.

———. "The Two Postwar Eras and the Conditions for Stability in Twentieth-Century Western Europe." *AHR* 86 (April 1981): 327–352.

Maizels, Alfred. *Industrial Growth and World Trade.* Cambridge: Cambridge University Press, 1963.

Manz, Mathias. "Stagnation und Aufschwung in der französischen Besatzungszone von 1945 bis 1948." Dissertation, Mannheim, 1968.

Marburg, Theodore F. "Government and Business in Germany: Public Policy toward Cartels." *BHR* 38 (Spring 1964): 78–101.

Mason, Timothy W. "The Primacy of Politics: Politics and Economics in National Socialist Germany." Pp. 165–195 of S. J. Woolf, ed., *The Nature of Fascism.* New York: Random House, 1968.

Messer, Robert L. *The End of an Alliance: James F. Byrnes, Roosevelt, Truman, and the Origins of the Cold War.* Chapel Hill, NC: University of North Carolina Press, 1982.

Metzner, Alfons. *Die chemische Industrie de Welt*, Bd. I: *Europa.* Düsseldorf: Econ, 1955.

Meyer, Fritz W. "Der Aussenhandel der westlichen Besatzungszonen Deutschlands und der Bundesrepublik 1945–1952." Pp. 258–285 of *Wirtschaft ohne Wunder.* Erlenbach-Zurich: Eugen Rentsch, 1953.

Milert, Werner. "Die verschenkte Kontrolle. Bestimmungsgründe und Grundzüge der britischen Kohlenpolitik im Ruhrbergbau 1945–1948." Pp. 105–119 of Dietmar Petzina and Walter Euchner, eds., *Wirtschaftspolitik im britischen Besatzungsgebiet 1945–1949.* Düsseldorf: Schwann, 1984.

Milward, Alan. *The German Economy at War.* London: The Athlone Press, 1965.

———. "Grossbritannien, Deutschland und der Wiederaufbau Westeuropas." Pp. 25–40 of Dietmar Petzina and Walter Euchner, eds., *Wirtschaftspolitik im britischen Besatzungsgebiet 1945–1949*. Düsseldorf: Schwann, 1984.

———. *The Reconstruction of Western Europe, 1945–1951*. London: Methuen, 1984.

Mioche, Philippe. "The Origins of the Monnet Plan: How a Transitory Experiment Responded to Deep-rooted Needs." European University Institute (Florence), EUI Working Paper No. 79 (1984).

Miscamble, Wilson. "Thurman Arnold Goes to Washington: A Look at Antitrust Policy in the later New Deal." *BHR* 56 (Spring 1982): 1–15.

Mommsen, Hans, and U. Borsdorf, eds. *Glück auf, Kameraden! Die Bergarbeiter und ihre Organisationen in Deutschland*. Köln: Bund-Verlag, 1979.

Morris, Peter J. T. "The Development of Acetylene Chemistry and Synthetic Rubber by I.G. Farbenindustrie Aktiengesellschaft, 1926–1945." Dissertation, Oxford, 1982.

Müller, Georg. *Die Grundlegung der westdeutschen Wirtschaftsordnung im Frankfurter Wirtschaftsrat 1947–1949*. Frankfurt: Haag und Herchen, 1982.

Nagel, Alfred von. "Methanol—Treibstoffe." *Schriftenreihe des Firmenarchivs der BASF*, Bd. 5 (1970).

Neebe, Reinhard. *Grossindustrie, Staat und NSDAP 1930–1933. Paul Silverberg und der Reichsverband der deutschen Industrie in der Krise der Weimarer Republik*. Göttingen: Vandenhoeck & Ruprecht, 1981.

Neumann, Franz. *Behemoth: The Structure and Practice of National Socialism*. (Reprinted of 2d ed. with new appendix from 1944.) New York: Octagon Books, 1963.

Nolte, Ernst. "Big Business and German Politics: A Comment." *AHR* 75 (October 1969): 71–78.

Nüske, Gerd F. "Neuere Literatur zur Geschichte der südwestdeutschen Länder 1945–1952." Pp. 383–422 of Hans Schwarzmaier, ed., *Landesgeschichte und Zeitgeschichte. Kriegsende und demokratischer Neubeginn am Oberrhein*. Karlsruhe: Kommissionsverlag G. Braun, 1980.

Overy, Richard J. "Heavy Industry and the State in Nazi Germany: The Reichswerke Crisis." *EHQ* 15 (July 1985): 313–340.

———. "Hitler's War and the German Economy: A Reinterpretation." *Economic History Review* 35 (May 1982): 272–291.

Papavassiliou, Nikolaos K. "Organisatorische Voraussetzungen für Expansionsstrategien multinationaler Unternehmungen. Mit einer empirischen Untersuchung deutscher multinationaler Unternehmungen der chemischen Industrie." Dissertation, Köln, 1977.

Paterson, Thomas, G. "The Quest for Peace and Prosperity: International Trade, Communism, and the Marshall Plan." Pp. 78–112 of Barton J. Bernstein, ed., *Politics and Policies of the Truman Administration*. Chica-

go: Quadrangle Books, 1970.
Peterson, Edward N. *The American Occupation of Germany: Retreat to Victory.* Detroit: Wayne State University Press, 1977.
———. "Eine Beurteilung der Einwirkung Amerikas auf Deutschland 1945–1952." Pp. 507–525 of Joachim Hütter and others, eds., *Tradition und Neubeginn. Internationale Forschungen zur deutschen Geschichte im 20. Jahrhundert.* Köln: Heymanns, 1975.
Petzina, Dietmar. *Autarkiepolitik im dritten Reich. Der nationalsozialistische Vierjahresplan.* Stuttgart: Deutsche Verlags-Anstalt, 1968.
———. "The Origins of the European Coal and Steel Community: Economic Forces and Political Interests." *ZgS* 137 (September 1981): 450–468.
———, and Walter Euchner, eds. *Wirtschaftspolitik im britischen Besatzungsgebiet 1945–1949.* Düsseldorf: Schwann, 1984.
Piettre, André. *L'économie allemande contemporaine (Allemagne occidentale), 1945–1952.* Paris: Editions M. Th. Génin, Librairie de Médicis, 1952.
Pingel, Falk. "Der aufhaltsame Aufschwung. Die Wirtschaftsplanung für die britische Zone im Rahmen der aussenpolitische Interessen der Besatzungsmacht." Pp. 41–64 in Dietmar Petzina and Walter Euchner, eds., *Wirtschaftspolitik im britischen Besatzungsgebiet 1945–1949.* Düsseldorf: Schwann, 1984.
———. "Politik deutscher Institutionen in den westlichen Besatzungszonen 1945–1948." *Neue politische Literatur* 25 (1980): 341–358.
Plumpe, Gottfried. "Industrie, technischer Fortschritt und Staat. Die Kautschuksynthese in Deutschland 1906–1944/45." *Geschichte und Gesellschaft* 9 (1983): 564–597.
———. "The I.G. Farben Group as a Multinational Enterprise between the Two World Wars." European University Institute (Florence), EUI Colloquium Papers (1984).
———. "Konzentrationsbewegung und Wiedereingliederung der deutschen Wirtschaft in ihre weltwirtschaftlichen Beziehungen nach dem Ersten Weltkrieg." Pp. 129–144 of Hans Pohl, ed., *Kartelle und Kartellgesetzgebung in Praxis und Rechtsprechung vom 19. Jahrhundert bis zur Gegenwart.* Stuttgart: Steiner, 1985.
Pohl, Manfred. *Wiederaufbau, Kunst und Technik der Finanzierung 1947–1953. Die ersten Jahre der Kreditanstalt für Wiederaufbau.* Frankfurt: Knapp, 1973.
Pohlenz, Michael. "Leverkusen und das Bayer-Werk in den Jahren 1944–1946." M.A. thesis, Köln, 1981.
Poidevin, Raymond. *L'allemagne et le monde au XXe siècle.* Paris: Masson, 1983.
———. "Frankreich und die Ruhrfrage 1945–1951." *Historische Zeitschrift* 228 (1979): 317–334.
———. "Die französische Deutschlandpolitik 1943–1949." Pp. 15–25 of

Claus Scharf and Hans-Jürgen Schröder, eds., *Die Deutschlandpolitik Frankreichs und die französische Zone 1945–1949*. Wiesbaden: Steiner, 1983.

Pritzkoleit, Kurt. *Bosse Banken Börsen. Herren über Geld und Wirtschaft*. Wien: Verlag Kurt Desch, 1954.

———. *Männer Mächte Monopole. Hinter den Türen der westdeutschen Wirtschaft*. Düsseldorf: Karl Rauck, 1953.

———. *Die neuen Herren. Die Mächtigen in Staat und Wirtschaft*. Wien: Verlag Kurt Desch, 1955.

Pross, Helge. *Manager und Aktionäre in Deutschland. Untersuchungen zum Verhältnis von Eigentum und Verfügungsmacht*. Frankfurt: Europäische Verlagsanstalt, 1965.

Prowe, Diethelm. "Economic Democracy in Post-World War II Germany: Corporatist Crisis Response 1945–1948." *Journal of Modern History* 57 (September 1985): 451–482.

Pünder, Tilmann. *Das bizonale Interregnum. Die Geschichte des Vereinigten Wirtschaftsgebietes 1946–1949*. Berlin: Grote, 1966.

Quilitzsch, Siegmar. *Die Rolle der Sowjetunion beim Neuaufbau der Chemieindustrie im Bitterfelder Gebiet 1945–1947. Am Beispiel der Filmfabrik Agfa-Wolfen*. Wolfen: Betreibsgruppe der DSF der VEB Filmfabrik Agfa-Wolfen, 1961.

Radandt, Hans. "Die I.G. Farbenindustrie A.G. und Südosteuropa bis 1938." *JfW* (Teil III, 1966): 146–185.

Rappaport, Armin. "The United States and European Integration: The First Phase." *Diplomatic History* 5 (Spring 1981): 121–149.

Räuschel, Jürgen. *Die BASF. Zur Anatomie eines multinationalen Konzerns*. Köln: Pahl-Rugenstein, 1975.

Reichelt, W.-O. *Das Erbe der I.G. Farben*. Düsseldorf: Econ, 1956.

Rich, Norman. *Hilter's War Aims: Ideology, the Nazi State, and the Course of Expansion*. Vol. I. New York: W. W. Norton, 1973.

Richter, Rudolf, ed. "Currency and Economic Reform. West Germany after World War II. A Symposium." Special issue of *ZgS* 135 (September 1979): 1–532.

———, and Wolfgang F. Stolper, eds. "Economic Reconstruction in Europe: The Reintegration of Western Germany. A Symposium." Special issue of *ZgS* 137 (September 1981): 341–663.

Ritschl, Albrecht. "Die Währungsreform von 1948 und der Wiederaufstieg der westdeutschen Industrie. Zu den Thesen von Mathias Manz und Werner Abelshauser über die Produktionswirkungen der Währungsreform." *VfZ* 33 (January 1985): 136–165.

Rode, Norbert. "Britische Besatzungspolitik in Niedersachsen 1945–1947." *Ergebnisse* 7 (1979): 5–86.

Schall, Horst. *Die chemische Industrie Deutschlands unter besonderer Berücksich-*

tigung der Standortsfrage. Nürnberg: Selbstverlag der wirtschaftsgeographischen Instituts der Hochschule für Wirtschafts- und Sozialwissenschaften, 1959.

———. "Ein Jahrhundert chemische Industrie. Die Entwicklung der chemischen Industrie in Deutschland und ihre heutige Stellung zur Mineralölindustrie." *Oel. Zeitschrift für die Mineralölwirtschaft* 3 (February 1965): 34–41.

Scharf, Claus, and Hans-Jürgen Schröder, eds. *Die Deutschlandpolitik Frankreichs und die französische Zone 1945–1949*. Wiesbaden: Steiner, 1983.

———. *Die Deutschlandpolitik Grossbritanniens und die britische Zone 1945–1949*. Wiesbaden: Steiner, 1979.

———. *Politische und ökonomische Stabilisierung Westdeutschlands. Fünf Beiträge zur Deutschlandpolitik der westlichen Alliierten*. Wiesbaden: Steiner, 1977.

Scherziger, Karl Aloys. *Bei I.G. Farben. Roman*. München: Andermann, 1953.

Schmidt, Eberhard. *Die verhinderte Neuordnung 1945–1952. Zur Auseinandersetzung um die Demokratisierung der Wirtschaft in den westlichen Besatzungszonen und in der Bundesrepublik Deutschland*. Frankfurt: Europäische Verlags-Anstalt, 1970.

Schmitt, Hans A. "The European Coal and Steel Community: Operations of the First European Antitrust Law, 1952–1958." *BHR* 38 (Spring 1964): 102–122.

———, ed. *U.S. Occupation in Europe after World War II*. Lawrence, Kansas: The Regents Press of Kansas, 1978.

Schneider, Ullrich. "Grundzüge britischer Deutschland- und Besatzungspolitik." *Zeitgeschichte* 19 (December 1981): 73–89.

Schreiber, Peter Wolfram. (Pseudonym of Autorenkollektiv von Mitgliedern der Kommunistischen Studentengruppen.) *I.G. Farben. Die unschuldigen Kriegsplaner*. Stuttgart: Verlag Neuer Weg, 1978.

Schreyer, Hermann. "Der I.G.-Farben-Konzern, seine Vorgänger und Nachfolger. Ein Beitrag zur Organisationsgeschichte der deutschen Chemieindustrie." *Archivmitteilungen* 16 (1966): 101–106, 148–158.

Schulte, Heinz. "Die britische Militärpolitik im besetzten Deutschland 1945–1949." *Militärgeschichtliche Mitteilungen* 31 (1982): 51–75.

Schwabe, Klaus. "Die amerikanischen Besatzungspolitik in Deutschland und die Entstehung des 'Kalten Krieges' (1945/46)." Pp. 311–332 of A. Fischer and others, eds., *Russland-Deutschland-Amerika. Festschrift für Fritz T. Epstein zum 80. Geburtstag*. Wiesbaden: Steiner, 1978.

Schwartz, Günther. "Die chemische Industrie als Wachstumsindustrie." Pp. 105–111 of Verband der Chemischen Industrie, ed., *Herrn Dr. Felix Ehrmann zum 60. Geburtstag*. Frankfurt: n.p., 1961.

Schwartz, Hans-Peter. *Vom Reich zur Bundesrepublik. Deutschland im Wider-*

streit der aussenpolitischen Konzeptionen in den Jahren des Besatzungsherrschaft 1945–1949. 2d enlarged ed. Stuttgart: Klett-Cotta, 1980.

Schwarzmaier, Hans, ed. *Landesgeschichte und Zeitgeschichte. Kriegsende und demokratischer Neubeginn am Oberrhein*. Karlsruhe: Kommissionsverlag G. Braun, 1980.

Schweitzer, Arthur. *Big Business in the Third Reich*. Bloomington, IN: Indiana University Press, 1964.

Senft, Helmuth. "Wirtschaftswissenschaftliche Probleme der Entflechtung. Ein Beitrag zur Entflechtungspolitik." Dissertation, Frankfurt, 1952.

Stamm, Thomas. *Zwischen Staat und Selbstverwaltung. Die deutsche Forschung im Wiederaufbau 1945–1965*. Köln: Verlag Wissenschaft und Politik, 1981.

Stein, Eberhard. "Die Entstehung der Leunawerke und die Anfänge der Arbeiterbewegung in den Leunawerken während des ersten Weltkrieges und der Novemberrevolution." Dissertation, Halle-Wittenberg, 1960.

Stocking, George, and Myron Watkins. *Cartels in Action: Case Studies in International Business Diplomacy*. New York: Twentieth Century Fund, 1947.

Stokes, Raymond G. "The Oil Industry in Nazi Germany, 1936–1945." *Business History Review* 59 (Summer 1985): 254–277.

———. "Germany Energy in the Postwar U.S. Economic Order, 1945–1951." *Journal of European Economic History*. Forthcoming.

Stolper, Gustav, Karl Häuser, and Knut Borchardt. *The German Economy 1870 to the Present*. New York: Harcourt, Brace & World, 1967.

Stratmann, Friedrich. *Chemie unter Zwang?*. Stuttgart: Franz Steiner, 1985.

———. "Stukturen der Bewirtschaftung in der Nachkriegszeit. Das Beispiel der Chemiebewirtschaftung in der britischen und der Bizone 1945 bis 1948." Pp. 153–172 of Dietmar Petzina and Walter Euchner, eds., *Wirtschaftspolitik im britischen Besatzungsgebiet 1945–1949*. Düsseldorf: Schwann, 1984.

Tammen, Helmuth. *Die I.G. Farbenindustrie A.G. (1925–1933)*. Berlin: Verlag H. Tammen, 1978.

Taylor, Graham D. "The Axis Replacement Program: Economic Warfare and the Chemical Industry in Latin America, 1942–1944." *Diplomatic History* 8 (Spring 1984): 145–164.

———. "The Rise and Fall of Antitrust in Occupied Germany, 1945–1948." *Prologue* 11 (1979): 23–39.

———, and Patricia E. Sudnick. *Du Pont and the International Chemical Industry*. Boston: Twayne Publishers, 1984.

Timm, B. "Wettbewerb zwischen Kohle und Erdöl bei der Rohstoffver-

sorgung der chemischen Industrie." *Chemie-Ingenieur-Technik* 40 (1968): 1-10.
Todd, Douglas. "Synthetic Rubber in the German War Economy: A Case of Economic Dependence." *Journal of European Economic History* 10 (Spring 1981): 153-165.
Treue, W. "Carl Duisbergs Denkschrift von 1915 zur Gründung der 'Kleinen I.G.'" *Tradition* 8 (1963): 193-227.
Turner, Henry A., Jr. *German Big Business and the Rise of Hitler.* New York: Oxford University Press, 1985.
Vietor, Richard H. K. *Energy Policy in America since 1945: A Study in Business-Government Relations.* Cambridge: Cambridge University Press, 1984.
Vogel, Walter. "Deutschland, Europa und die Umgestaltung der amerikanischen Sicherheitspolitik 1945-1949." *VfZ* 19 (January 1971): 64-82.
———. *Westdeutschland 1945-1950. Der Aufbau von Verfassungsund Verwaltungseinrichtungen über die Länder der drei westlichen Besatzungszonen.* 3 vols. Boppard: Harald Boldt Verlag, 1956-1983.
Wandel, Eckhard. *Die Entstehung der Bank deutscher Länder und die deutsche Währungsreform 1948. Die Rekonstruktion des westdeutschen Geld- und Währungssystems 1945-1949 unter Berücksichtigung der amerikanischen Besatzungspolitik.* Frankfurt: Fritz Knapp Verlag, 1980.
Wiel, Paul. *Wirtschaftsgeschichte des Ruhrgebietes. Tatsachen und Zahlen.* Essen: Siedlungsverband Ruhrkohlenbezirk, 1970.
Willis, F. Roy. *France, Germany and the New Europe, 1945-1967.* Rev. and exp. ed. Stanford: Stanford University Press, 1968.
———. *The French in Germany, 1945-1949.* Stanford: Stanford University Press, 1962.
———, ed. *European Integration.* New York: New Viewpoints, 1975.
Winkel, Harald. *Die Wirtschaft im geteilten Deutschland 1945-1970.* Wiesbaden: Steiner, 1974.
Winkler, Heinrich August, ed. *Politsche Weichenstellung im Nachkriegsdeutschland 1945-1953.* Göttingen: Vandenhoeck & Ruprecht, 1979.
Winnacker, Karl, and Leopold Küchler, eds. *Chemische Technologie.* 3d rev. ed. 7 vols. München: Carl Hansen, 1970.
Yates, P. Lamartine. *Forty Years of Foreign Trade: A Statistical Handbook with Special Reference to Primary Products and Underdeveloped Countries.* London: George Allen & Unwin, 1959.
Yergin, Daniel. *Shattered Peace: The Origins of the Cold War and the National Security State.* Boston: Houghton-Mifflin, 1978.
Zank, Wolfgang. "Wirtschaftsplanung und Bewirtschaftung in der Sowjetischen Besatzungszone—Besonderheiten und Parallelen im Vergleich

zum westlichen Besatzungsgebiet 1945–1949." *Vierteljahrshefte für Sozial- und Wirtschaftsgeschichte* 71 (1984): 485–504.

Zapf, Wolfgang. "Die deutschen Manager. Sozialprofil und Karrierweg." Pp. 136–149 of Zapf, ed., *Beiträge zur Analysen der deutschen Oberschicht*. 2d exp. ed. München: R. Piper & Co., 1965.

Ziebura, Gilbert. *Die deutsch-französischen Beziehungen seit 1945. Mythen und Realitäten*. Pfuldingen: Günther Neske, 1970.

Ziemke, Earl. *The U.S. Army in the Occupation of Germany 1944–19846*. Washington: Center of Military History, U.S. Army, 1975.

Zink, Harold. *The United States in Germany, 1944–1955*. Princeton, NJ: D. Van Nostrand, 1957.

INDEX

Abs, Hermann, 176
Adenauer, Konrad, 171–172, 175
AEG, 157
Agfa, 7, 12, 176, 179–180
Allied Control Commission Law 9 (Breakup of I.G.), 118, 156, 174
Allied High Commission, 171, 172–173, 178; AHC Law 25 (Control of research), 61; AHC Law 35 (Breakup of I.G.), 175
American Cyanamide, 136
Amick, Erwin H., 176
Antitrust Division (of U.S. Department of Justice), 53, 56; role in developing and implementing U.S. policy toward I.G., 41–48, 153–154, 177–178
Arnold, Thurman, 43
Association of British Chemical Manufacturers (ACBM), 80, 81
Attlee, Clement, 76
Auschwitz, 19, 21, 22, 28–29, 30, 45, 54, 73, 152, 153
Autarky, 2, 27, 29–30, 93, 202–203

Bain, Frederick, 79
Balke, Stefan, 154–155
BASF, 2, 4, 7, 11, 36, 39, 49, 53, 73, 91, 98, 105, 119–120, 162, 164, 174, 178–179, 197, 202, 203–204, 205; capsule history, 86–91; causes and effects of 1948 explosion, 97–101; effects of Nuremberg Trial on, 155; French control of, 91–103; in I.G. period, 17–23 passim; production performance from 1946, 160–161, 167–169; refounding of, 176–177, 180, 181, 182, 183–184, 189; size compared to French chemical industry, 91
Bayer, 4, 7, 31, 36, 41, 49, 53, 54, 68, 91, 98, 105, 156, 164, 174, 178–179, 191, 195–196, 197, 203–205; British policy on , 83–85; capsule history, 70–75; coal supply to, 130–134; in I.G. period, 17–23 passim; preparations for and effects of currency reform, 146–149; production performance of, 116–117, 160–161, 167–169; refounding of, 176–177, 179–180, 181–182, 183, 184, 189

Behringwerke (Marburg), 50, 51 n. 50, 179, 181, 191
Bernstein, Bernard, 44, 153
Bevin, Ernest, 76, 109
Bipartite Control Office (BICO), 120–121
Bipartite I.G. Farben Control Office (BIFCO), 121, 125, 156, 159, 174–175
Bizonal Economic Control Group (BECG), 120–121, 128–129, 134
Bizone, 38, 82, 95, 103, 105; American cast to policy in, 108, 109, 122, 170, 177; chemical control in, 120–125; formation of, 108, 109–110; growing French cooperation with, 159–160, 170–171; policy on I.G. Farben, 156–159
Board of Trade (BOT, British), 79, 80, 81
Bobingen, 17, 176
Böhringer, 14
Borkin, Joseph, 3, 43, 152–153
Bosch, Carl, 8, 13, 16, 23, 51, 89, 90
Brabag, 157
Brearley, George, 176
Brecht, Gustav, 157, 176
British Rayon Federation, 79, 81
British zone of occupation: chemical industry in, 70–71; composition of, 64–65; control of chemical industry in, 67–70; policy formation, 65–67; policy in context, 75–82; policy on I.G. Farben, 82–85; socialization, 77
Broadbent, H. L., 68
Brunck, Heinrich, von, 88–89, 90
Brüning, Adolph, 49–50
Brunsbüttel, 204
Bücher, Hermann, 157
Burgess, A. L., 191
Burghartz, Arnold, 158
Bütefisch, Heinrich, 23–24, 28, 29, 152, 157
Byrnes, James, 109

Casella, 8, 12, 50, 51 n. 50, 176, 181
CCG, BE (Control Council for Germany, British Element), 67, 80
Chemical industry, German: in British zone, 70–71; coal allocation to, 127–135; cooperation with European and

287

U.S. firms, 197–198; early history, 4–11; effects of currency reform on, 140–149; in French zone, 86–87; investment in, 184–189; write-offs and, 186–187; Marshall Plan and, 149–151; place of I.G. in, 11–14; production performance, 110–117, 137–145, 160–163, 164–169; reestablishment of foreign trade, 194–196; undercapitalization of, 187–189; in U.S. zone, 37–38; wartime damage to factories, 33, 52, 73–76 passim, 90–91
Chemische Werke Albert, 60
Churchill, Winston 39, 76
Ciba-Geigy, 91
Clay, General Lucius, 46, 193–194
Coal supply, 95, 110, 163; to chemical industry in Bizone, 115–116, 117; German discussions on, 126–135
Combined Area. *See* Bizone
Coster, Harry D., 61, 62
Counterpart funds, 150–151. *See also* European Recovery Program
Courtauld, S., 79, 80
Cripps, Stafford, 80
Currency reform, 95, 106, 125, 137, 139, 159, 170–171, 186–187; effects of on major I.G. successors, 147–149, 160–163; effects on chemical industry, 144–149; preparations for by chemical firms, 146–147
Curtis, Francis, 75

Decartelization. See *Entflechtung*
Dencker, Paul, 148, 158–159
Denivelle, Leon C., 176
Deutsche Solvay-Werke, 14, 133
Dieter, Henning B., 115–116
Dillon, Read and Co., 46
Dormagen, 71, 72–73, 76, 83, 98, 176, 179, 204
Draper, William, 46–48
Dreibund, 8, 10, 12
Dreiverband, 8, 10, 12, 50
Dresdner Bank (Rhein-Ruhr Bank), 181
DuBois, Josiah, 54–55, 153
Duerholt, Dr., 126
Duisberg, Carl, 7–11, 13, 16, 51, 71–72, 73, 90, 157
Duisburger Kupferhütte, 71
Dunlop, 68
Du Pont, 9, 13, 43, 204–205
Dürrfeld, Walter, 152
Dynamit A.G., 133

Economic miracle, xxiv; chemical industry and, 137–151
Eden, Anthony, 76
Ehrlich, Paul, 50
EIPS (British Economic and Industrial Planning Staff), 65–67
Eisenhower, Dwight D., 39
Elberfeld, 71, 72–73, 76, 83, 98, 179
Engelhorn, Friedrich, 4, 87
Entflechtung, 53, 93, 134, 161–163, 191; bizonal policy, 120–121, 156–159; establishment of I.G. successors, 179–189; four-power policy, 118–120; process of, 173–179; *Selbstentflechtung* plans during war, 31–33; U.S. attitudes toward, 44–48
Erhard, Ludwig, 139, 175
European Coal and Steel Community (ECSC), xxiii, 172, 196–197, 198
European Recovery Program (ERP), xxiii, 137, 139–140, 159, 162, 185–186; role in resurgence of chemical industry, 149–151

FARDIP (I.G. Farben Dispersal Panel), 125, 156–159, 160, 161–162, 175–176
Fogler, Mayor, 75
Forster, Karl-Heinz, 186–187
Fouchier, J. P., 99
Francolor S.A., 97, 192
French zone of occupation: ambivalence in policy, 94, 103–105; chemical industry in, 86–87; composition of, 86; control of I.G., 94–101; policy toward I.G. Farben, 91–94
Frowein, Friedrich, 61–62
Funk, Walther, 24

Gajewski, Fritz, 156 n. 34
Gattineau, Heinrich, 23–24, 152, 156
General Motors, 13
Gershofen, 50
Gewerkschaft Victor Chemische Werke, 203
Griesheim, Chemische Fabrik, 12, 50, 51 n. 50, 179, 195

Haberland, Ulrich, 31–32, 54, 58, 83–85, 132–133, 146–148, 155, 158, 164, 174, 183
Hansen, Kurt, 195–196
Hauck, Dr., 123, 132–133
Heintzeler, Wolfgang, 183
Hercules Powder Co., 136
Heubaum, Dr., 126

Index

Heyde, Erich, von der, 152
Heydebreck, 19, 20, 22, 104
Himmler, Heinrich, 28, 29
Hitler, Adolph, 2, 24, 29, 74, 203
Hoechst, 4, 8, 12, 36, 38, 47–48, 51 n. 50, 62, 63, 97, 164, 174, 178–179, 195, 197, 201, 203–205; effects of currency reform on, 147–149; in I.G. period, 17–23 passim; production performance of, 116–117, 160–161, 167–169; refounding of, 176–177, 179, 181, 182–183, 184, 189; supply shortages, 115, 125–126, 133; under U.S. control, 49–57, 60–61
Hüls, 12, 19, 20, 22, 133

ICI (Imperial Chemical Industries), 13, 67–68, 78–79
I.G. Farben Control Committee (COIG), 118–120
I.G. Farben Control Office, 56–57, 58, 68, 118–119, 159–160, 178
I.G. Farbenindustrie A.G.: in Abwicklung, 177; Allied control of, 118–120; Berlin Works Group, 17–21 passim; bizonal control of, 120–122; Central Germany Works Group, 17–21 passim, 51; differences among works groups, 17–23; formation of, 10–11; Lower Rhine Group (*see* Bayer); Maingau Group (*see* Hoechst); organization, 14–18; planning for postwar period, 30–33; profits and dividends, 25–27, 188; relations with Nazi regime, 2, 23–33, 89; research expenditures, 13, 25–27; size, 11–14; technical connections between factories, 18–19, 50–51, 72–73, 88; Upper Rhine Group (*see* BASF). See also *Entflechtung*
I.G. Farben Liquidation Committee, 159, 180–181, 188
Ilgner, Max, 24, 156 n. 34
Indanthren Warenzeichenverband, 192
International Authority on the Ruhr, 173
IPCOG (U.S. Informal Policy Committee on Germany), 39–40, 66

JCS 1067 (occupation directive), 40–41, 42, 66
JEIA (Joint Export-Import Agency), 193

Kali-Chemie, 133
Kalle & Co., 4, 12, 50, 52
Keeling, R., 81

Keiser, Dr., 130, 132–133
Kelleher, Grant, 43
Keppler, Wilhelm, 28
Knapsack, 50, 54, 71, 85, 125–126, 133, 179
Knoll A.G., 204
Koch, Robert, 50
Koenig, Pierre, 99, 118
Krauch, Carl, 16, 27–28, 29, 54, 61, 151–152
Kreikamp, H. D., 177–179
Krekeler, Heinz, 104–105

Labour Party, 76–77
Länderfachausschuss Chemie (LFA), 124, 126, 128–129, 133
Landeswirtschaftsamt (LWA, or Provincial Economic Office), 58–60, 115, 126, 201
Lanz, Kurt, 52, 184, 195, 196
Lefebure, Victor, 3
Leuna Works, 3, 17, 20, 22, 28, 29, 85, 88, 174
Liebig, Justus, von, 4
"Little I.G.," 10–11, 12, 50
Loehr, Oskar, 130–131, 158
Lucius, Eugen, 49
Ludewig, Walter, 183

McNarney, J. T., 118
Maltzan, von, Dr., 132
Mann, Wilhelm Rudolph, 24
Marshall Plan. *See* European Recovery Program
Martin, James S., 43, 46–47, 56
Matthes & Weber, 133
Maupin, Myron, 43, 160
ter Meer, Fritz, 152, 155–156
Meister, Wilhelm, 49
Menne, W. A., 176
Merck (German), 14
Merck & Co. (U.S.), 197
Miles Laboratories, 204
Military Security Board, 171 n. 7
Mills, Percy, 67
Milward, Alan, 86
Moehn, Eugen, 158
Moldenhauer, Paul, 125 n. 25
Monsanto, 204
Montecatini, 91
Montgomery, B. L., 118
Morgenthau, Henry, Jr., xxiii, 39
Moscow Foreign Ministers Conference, 105
Moulton, Herbert G., 52

Mueller, Rudolph, 61
Müller, August, 49

Newman, Randolph, 43
Nuremberg Trials, 25, 43, 44, 54–55, 85, 151–156

OEEC (Organization for European Economic Cooperation), 159, 172
Offenbach, 50, 51 n. 50, 179
OMGH (Office of Military Government, U.S. for Hesse), 61, 115, 190; confused control of I.G. Farben, 48–49; organization, 55–56, 59–60
OMGUS (Office of Military Government, U.S.), 61, 62, 191; organization, 55–56
Ott, Emil, 136

Patents and trademarks, 5, 6, 41–42, 61, 190–194
Petersberg Agreement, 172–173
Pohland, Dr., 122, 123, 132
Potsdam Conference, 76
Prentzel, Felix, 175–176, 180
Pross, Helge, 188
Pünder, Hermann, 193–194

Reich Patent Law, 6
Reppe, Walter, 183
Research, 2, 13, 25–27, 36, 62, 72, 87–88, 89–90, 207
Ritter, Egon, von, 158
Robinson, C. S., 67–68
Roosevelt, Franklin D., 39
Rospatt, Heinrich, von, 175–176, 180
Rottweil, 176
Ruhrchemie A.G., 204

Sammet, Rolf, 184
Scheffer, R., 102
Schenley Laboratories, 197
Schkopau, 19, 20, 22
Schlotterer, Gustav, 30
Schmitz, Hermann, 32–33
Schneider, Christian, 152
Schuman, Robert, 172
Seal, Eric, 80
Shell, Royal Dutch, 68, 197
Shepard, Norman A., 136
Spaatz, Carl A., 33

Stalin, Josef, 76
Standard Oil of New Jersey, 13, 43

Taillefer, A., 102
Tammen, Helmuth, 17
Technical exploitation, 41–42, 47, 78
Theurer, Dr., 122, 123
Timm, Bernhard, 183
Treasury Department (U.S.), 153–154; role in control of I.G., 43–48, 178; role in occupation planning, 39
Troutbeck, J. M., 65
Truman, Harry S., 39, 76
Turner, R. M. C. (Mark), 65, 67

Uerdingen (Farbenfabriken ter Meer), 4, 71, 72–73, 76, 83, 84, 85, 98, 204
U.S. Steel, 13
U.S. zone of occupation: chemical industry in, 37–38; composition of, 37; German impact on policy, 58–62; planning for occupation of, 38–40; U.S. policy on I.G. Farben, 42–49, 53–57, 63

VAW (Verwaltungsamt für Wirtschaft), later VfW (Verwaltung für Wirtschaft), 121–125, 127–135, 139

Wacker, Dr. Alexander, 48, 133
War damage, 33, 185, 201; at BASF factories, 90–91; at Bayer factories, 73–76 passim; at Hoechst factories, 52
Warnecke, Dr., 131
WASAG, 32, 156
Wickel, Helmuth, 3
Wilhelmi, Lothar, 60–61
Willis, F. Roy, 105
Winnacker, Karl, 51, 54, 85, 180–181, 183, 184
Wintershall A.G., 203
Wirtschaftswunder. *See* Economic miracle
Wohltat, Hermann, 176
Wolfen-Bitterfeld, 17, 20–21, 22, 51
Wood, Derek, 81
Wurster, Karl, 155, 161, 183

Zhukov, Marshal G., 118
Zvegintzow, M., 68